MATH TRICKS, BRAIN TWISTERS, AND PUZZLES

MATH TRICKS,
BRAIN TWISTERS,
AND PUZZLES

by

JOSEPH DEGRAZIA, Ph.D.

Illustrated by

ARTHUR M. KRAFT

BELL PUBLISHING COMPANY
NEW YORK

This book was previously titled *Math Is Fun*.

This edition is published by Bell Publishing Company,
a division of Crown Publishers, Inc.,
by arrangement with Emerson Books, Inc.
a b c d e f g h
BELL 1981 EDITION

Manufactured in the United States of America

Library of Congress Cataloging in Publication Data
Degrazia, Joseph, 1883-
    Math tricks, brain twisters, and puzzles.

    Earlier ed. published under title: Math is fun.
    1. Mathematical recreations. 2. Mathematics—
Problems, exercises, etc. I. Title.
QA95.D36 1981   793.7'4   80-26941
ISBN 0-517-33649-9

CONTENTS

PREFACE

This book is the result of twenty years of puzzle collecting. For these many years I have endeavored to gather everything belonging to the realm of mathematical entertainment from all available sources. As an editor of newspaper columns on scientific entertainment, I found my readers keenly interested in this kind of pastime, and these readers proved to be among my best sources for all sorts of problems, both elementary and intricate.

Puzzles seem to have beguiled men in every civilization, and the staples of scientific entertainment are certain historic problems which have perplexed and diverted men for centuries. Besides a number of these, this book contains many problems never before published. Indeed, the majority of the problems have been devised by me or have been developed out of suggestions from readers or friends.

This book represents only a relatively small selection from an inexhaustible reservoir of material. Its purpose is to satisfy not only mathematically educated and gifted readers but also those who are on less good terms with mathematics but consider cudgeling their brains a useful pastime. Many puzzles are therefore included, especially in the first chapters, which do not require even a pencil for their solution, let alone algebraic formulas. The majority of the problems chosen, however, will appeal to the puzzle lover who has not yet forgotten the elements of arithmetic he learned in high school. And finally, those who really enjoy the beauties of mathematics will find plenty of problems to rack their brains and test their knowledge and ingenuity in such chapters as, for example, "Whimsical Numbers" and "Playing with Squares".

The puzzles in this book are classified into groups so that the reader with pronounced tastes may easily find his meat. Those familiar with mathematical entertainment may miss certain all-too-well-known types, such as the famous magic squares. I believe, however, that branches of mathematical entertainment which have long since developed into special sciences belong only in books that set out to treat them exhaustively. Here we must pass them by, if only for reasons of space. Nor have geometrical problems been included.

Lack of space has also made it impossible to present every solution fully. In a great many instances, every step of reasoning, mathematical and other, is shown; in others, only the major steps are indicated; in others still, just the results are given. But in every single class of problems, enough detailed solutions are developed and enough hints and clues offered to show the reader his way when he comes to grips with those problems for which only answers are given without proof. I hope that with the publication of this book I have attained two objectives: to provide friends of mathematics with many hours of entertainment, and to help some of the myriads who since their school days have been dismayed by everything mathematical to overcome their horror of figures.

I also take this opportunity of thanking Mr. André Lion for the valuable help he has extended me in the compilation of the book.

Joseph Degrazia, Ph.D.

CHAPTER I

TRIFLES

We shall begin with some tricky little puzzles which are just on the borderline between serious problems and obvious jokes. The mathematically inclined reader may perhaps frown on such trifles, but he should not be too lofty about them, for he may very well fall into a trap just because he relies too much on his arithmetic. On the other hand, these puzzles do not depend exclusively on the reader's simplicity. The idea is not just to pull his leg, but to tempt him mentally into a blind alley unless he watches out.

A typical example of this class of puzzle is the *Search for the missing dollar*, a problem—if you choose to call it one—which some acute mind contrived some years ago and which since then has traveled around the world in the trappings of practically every currency.

A traveling salesman who had spent several nights in a little upstate New York hotel asks for his bill. It amounts to $30 which he, being a trusting soul, pays without more ado. Right after the guest has left the house for the railroad station the desk clerk realizes that he had overcharged his guest $5. So he sends the bellboy to the station to refund the overcharge to the guest. The bellhop, it turns out, is far less honest than his supervisor. He argues: "If I pay that fellow only $3 back he will still be overjoyed at getting something he never expected—and I'll be richer by $2. And that's what he did.

Now the question is: If the guest gets a refund of $3 he had paid $27 to the hotel all told. The dishonest bellhop has kept $2. That adds up to $29. But this monetary transaction started with $30 being paid to the desk clerk. Where is the 30th dollar?

9

Unless you realize that the question is misleading you will search in vain for the missing dollar which, in reality, isn't missing at all. To clear up the mess you do not have to be a certified public accountant, though a little bookkeeping knowledge will do no harm. This is the way the bookkeeper would proceed: The desk clerk received $30 minus $5, that is, $25; the bellhop kept $2; that is altogether $27 on one side of the ledger. On the other side are the expenses of the guest, namely $30 minus $3, also equalling $27. So there is no deficit from the bookkeeper's angle, and no dollar is missing. Of course, if you mix up receipts and expenses and add the guest's expenses of $27 to the dishonest bellhop's profit of $2, you end up with a sum of $29, and a misleading question.

The following are further such puzzles which combine a little arithmetic with a dose of fun.

1. How much is the bottle?

Rich Mr. Vanderford buys a bottle of very old French brandy in a liquor store. The price is $45. When the store owner hands him the wrapped bottle he asks Mr. Vanderford to do him a favor. He would like to have the old bottle back to put on display in his window, and he would be willing to pay for the empty bottle. "How much?" asks Mr. Vanderford. "Well," the store owner answers, "the full bottle costs $45 and the brandy costs $40 more than the empty bottle. Therefore, the empty bottle is . . ." "Five dollars," interrupts Mr. Vanderford, who, having made a lot of money, thinks he knows his figures better than anybody else. "Sorry, sir, you can't figure," says the liquor dealer and he was right. Why?

2. Bad day on the used-car market.

A used-car dealer complains to his friend that today has been a bad day. He has sold two cars, he tells his friend, for $750 each. One of the sales yielded him a 25 per cent profit. On the other one he took a loss of 25 per cent. "What are you worrying about?" asks his friend. "You had no loss whatsoever."

"On the contrary, a substantial one," answers the car dealer. Who was right?

3. The miller's fee.

In a Tennessee mountain community the miller retains as his fee one-tenth of the corn the mountaineer farmers deliver for grinding. How much corn must a farmer deliver to get 100 pounds of cornmeal back, provided there is no loss?

4. Two watches that need adjusting.

Charley and Sam were to meet at the railroad station to make the 8 o'clock train. Charley thinks his watch is 25 minutes fast, while in fact it is 10 minutes slow. Sam thinks his watch is 10 minutes slow, while in reality is has gained 5 minutes. Now what is going to happen if both, relying on their timepieces, try to be at the station 5 minutes before the train leaves?

5. Involved family relations.

A boy says, "I have as many brothers as sisters." His sister says, "I have twice as many brothers as sisters." How many brothers and sisters are there in this family?

6. An ancient problem concerning snails.

You may have come across the ancient problem of the snail which, endeavoring to attain a certain height, manages during the day to come somewhat closer to its objective, though at a snail's pace, while at night, unfortunately, it slips back, though not all the way. The question, of course, is how long will it take the persevering snail to reach its goal? The problem seems to have turned up for the first time in an arithmetic textbook written by Christoff Rudolf and published in Nuremberg in 1561. We may put it this way (without being sure whether we do justice to the snail's abilities):

A snail is at the bottom of a well 20 yards deep. Every day it climbs 7 yards and every night it slides back 2 yards. In how many days will it be out of the well?

7. Cobblestones and water level.

A boat is carrying cobblestones on a small lake. The boat capsizes and the cobblestones drop to the bottom of the lake. The boat, being empty, now displaces less water than when fully loaded. The question is: Will the lake's water level rise or drop because of the cobblestones on its bottom?

8. Two gear wheels.

If we have two gear wheels of the same size, one of which rotates once around the other, which is stationary, how often will the first one turn around its own axle?

9. Divisibility by 3.

Stop a minute and try to remember how to find out quickly whether a number is divisible by 3. Now, the question is: Can the number eleven thousand eleven hundred and eleven be divided by 3?

10. Of cats and mice.

If 5 cats can catch 5 mice in 5 minutes, how many cats are required to catch 100 mice in 100 minutes?

11. Mileage on a phonograph record.

A phonograph record has a total diameter of 12 inches. The recording itself leaves an outer margin of an inch; the diameter of the unused center of the record is 4 inches. There are an average of 90 grooves to the inch. How far does the needle travel when the record is played?

CHAPTER II

On the Borderline of Mathematics

Here we have some puzzles on the borderline between arithmetic and riddle. Their solution hardly requires any knowledge of algebra though it does demand some logical reasoning and mental dexterity. In trying to solve puzzles like these, a person who knows his mathematics well has little advantage over the amateur arithmetician. On the contrary, he may often be at a disadvantage when he tries to use theories and a fountain pen to solve problems which require intuition and logical thinking rather than mathematical equations.

The first puzzle of this series is a distant relative of the ancient fallacy resulting from the statement that all Cretans are liars. Therefore, if a Cretan states that all Cretans are liars, he himself lies. Consequently, Cretans are not liars. That proves that the Cretan's statement was corrrect, namely that all Cretans are liars, and so on, endlessly.

12. Lies at dawn.

Let's presume there is such an island, not slandered Crete, but one which is inhabited by two tribes, the Atans and the Betans. The Atans are known all over the world to be inveterate liars, while a Betan always tells the truth. During one stormy night, a ship has run aground near the island. At dawn a man from the ship approaches the island in a rowboat and in the mist sees a group of three men. Knowing the bad reputation of one of the two tribes he wants to find out which of the two he will have to deal with. So he addresses the first man on the shore and asks him whether he is an Atan or a Betan. The man's answer is lost in the roaring of the breakers. However, the man in the boat understands what the second man yells across the surf: "He says he is a Betan. He is one and so am I." Then the third man points at the first and yells: "That isn't true. He is an Atan and I am a Betan." Now, what is the truth?

13. Quick decision during an air raid alarm.

This is supposed to have happened during an air raid alarm in London during the war. The sirens screamed; all lights go out. A man jumps out of bed and starts running for the air raid shelter. Then he realizes that it will be cold in the basement and that he had better get a pair of socks. He turns back and opens the drawer which holds his stockings. Only then

is he aware that all the lights are out, which makes it impossible for him to see what he grabs in the drawer. He is a man of self-respect and order and would hate to be seen with two socks of different colors. He knows that the drawer contains nothing but black and brown socks which, unfortunately, he cannot distinguish in the darkness. Now, what is the minimum number of socks the man must grab to be sure he will find a matched pair among them when he reaches the shelter?

14. Who is the smartest?

Each of three friends thinks that he is the smartest? To find out who really is, a fourth friend makes the following suggestion: He will paint on each of the three men's foreheads either a black or a white spot without any of the men knowing which color adorns his own brow. After each man has been marked, all three will be simultaneously led into the same room. Each one who sees a black spot on the forehead of one or two of his friends is supposed to raise his right hand. The one who finds out first whether he himself is marked black or white and is able to prove his statement will be recognized as the smartest of the three.

The referee now marks each of his three friends with a black spot and lets them enter the room simultaneously. As each of them sees two black spots, all three raise their hands. After a moment's hesitation one of them states: "I have a black spot." How did he reason this out?

15. Five Hats.

At a party, four people played a game. Three of them sat one behind the other so that Abe saw Bill and Cal, and Bill saw only Cal who sat in front and saw nobody. Dave had five hats which he showed to his three friends. Three of the hats

were blue, two were red. Now Dave placed a hat on the head of each of his three friends, putting aside the remaining two hats. Then he asked Abe what color his hat was. Abe said he couldn't tell. Bill, asked the color of his own hat, didn't know for certain either. Cal, however though he couldn't see any hat at all, gave the right answer when asked what color his hat was.

Do you know what color Cal's hat was and how he reasoned to find the correct answer?

16. "Unshakable testimony."

Six people, let's call them *A, B, C, D, E* and *F*, have witnessed a burglary and are only too willing to let the police know what the burglar—who, by the way, managed to escape—looked like. But you know how eyewitnesses' accounts go; the descriptions of the criminal differed in every important point, particularly with regard to the color of his hair and eyes, the color of his suit and probable age.

This is the testimony the police got from these six witnesses:

	Question I: (Hair)	Question II: (Eyes)	Question III: (Suit)	Question IV: (Age)
A	brown	blue	grey	34
B	blond	black	dark blue	30
C	red	brown	dark brown	34
D	black	blue	not dark brown	30
E	brown	black	grey	28
F	blond	brown	dark blue	32

Though these contradictory reports weren't much help, the police finally got their man and compared his real appearance with the six descriptions. They found that each of the six witnesses had made three erroneous statements and that each of the four questions had been answered correctly at least once. What did the burglar really look like?

17. The watchmen's schedule.

Four men, Jake, Dick, Fred and Eddie, are watchmen in a small factory in Hoboken. Their job consists of two daily shifts of six hours each, interrupted by a rest of several hours. Two of the men have to be on the job all the time; however, they are not supposed to be relieved at the same time. Each shift has to begin on the hour. Except for these rules, the four men may agree among themselves on any schedule they want.

So they have a meeting at which they ask for the following privileges: Jake wants to start his first shift at midnight and would like to be off by 4 p.m. Dick would like to be off from 10 a.m. to 4 p.m. Fred wants to relieve Dick after the latter's night shift. Eddie has to be on the premises at 9 a.m. to receive special instructions. After some hours of trial and error the four watchmen finally succeed in making out a shift schedule in accord with the regulations as well as their special wishes. What was this schedule?

What are their names?

In this group of puzzles the names and other characterisitcs of all the people mentioned have to be found with the aid of a number of clues. Most of these problems are further alike in that the various impossibilities have to be excluded by systematic elimination, so that finally the remaining possibilities become certainties. Let us start with a comparatively simple problem of this class as an introduction to more difficult ones and eventually to some really tough nuts to crack.

18. Cops and robbers.

There are communities where the same family names occur time and again. In one such community it happened that one day there were ten men at the police station, six of them named Miller. Altogether there were six policemen and four burglers. One Miller had arrested a Miller and one Smith a Smith. However, this burglar, Smith, was not arrested by his own brother. Nobody remembers who arrested Kelly but anyway, only a Miller or a Smith could have been responsible for that act. What are the names of these ten people?

19. Who is married to whom?

The following problem is more difficult than the preceding one. To give you a lead for the solution of still tougher ones we shall show you how this one should be solved by successive eliminations.

Imagine seven married couples meeting at a party in New York. All the men are employees of the United Nations, and so is a fifteenth male guest who happens to be single. Since these seven couples come from all four corners of the world, the fifteenth guest has never met any of them and so he does not know which of the women is married to which of the men. But as this man has the reputation of being outstandingly brainy, the seven couples propose not to introduce themselves formally to him but to let him find out by himself who is married to whom. All they tell him is the surnames of the men, Wade, Ford, Vitta, Storace, Bassard, Hard and Twist, and the first names of the women, Gertrude, Felicia, Lucy, Cecile, Maria, Olga and Charlotte.

The intelligent guest, No. 15, gets only two clues which are sufficient for him to conclude which of the men and women belong together. Everything else is left to his observations and power of reasoning. The two clues are: None of the men will dance with his own wife and no couple will take part in the same game.

This is what the clever guest observes and keeps in mind: Wade dances with Felicia and Cecile, Hard with Maria and Cecile, Twist with Felicia and Olga, Vitta with Felicia, Storace with Cecile and Bassard with Olga.

When they play bridge, first Vitta and Wade play with Olga and Charlotte. Then the two men are relieved by Storace and Hard while the two girls keep on playing. Finally, these two gentlemen stay in and play with Gertrude and Felicia.

These observations are sufficient for the smart guest to deduce who is married to whom. Are you, too, smart enough to conclude from these premises the surnames of the seven women present?

In case this problem proves a little tough, this is the way you should proceed (and the procedure is practically the same for all such problems): You mark off a set of names, noting, in orderly fashion, the negative facts, that is, all the combinations which have to be eliminated. In this case the horizontal spaces may represent the family names, the vertical columns the first names of the women. Since Wade danced with Felicia, her last name cannot be Wade, so you put an x where these two names cross. After you have marked all other negative

19

combinations with an x, your diagram will look like this:

	Felicia	Gertrude	Lucy	Cecile	Maria	Olga	Charlotte
Wade	x			x		x	x
Ford							
Vitta	x					x	x
Storace	x	x		x		x	x
Bassard						x	
Hard	x	x		x	x	x	x
Twist	x					x	

The rest is simple. You see immediately that only Lucy can be Mrs. Hard and that Olga must be Mrs. Ford. Having thus found a wife for Mr. Ford and a husband for Lucy we may put x's in all the fields of the second horizontal line and of the third vertical column (remember, an x stands for an impossibility). We now discover that Felicia can only be Mrs. Bassard, and we may now put x's in the empty spaces of the fifth line. This leaves another opening in the right bottom corner, indicating without doubt that Charlotte is Mrs. Twist. Proceeding similarly, we find quickly that Maria is Mrs. Storace, Cecile—Mrs. Vitta and Gertrude—Mrs. Wade.

20. The tennis match.

Six couples took part in a tennis match. Their names were Howard, Kress, McLean, Randolph, Lewis and Rust. The first names of their wives were Margaret, Susan, Laura, Diana, Grace and Virginia. Each of the ladies hailed from a different city, Fort Worth, Texas, Wichita, Kansas, Mt. Vernon, New York, Boston, Mass., Dayton, Ohio and Kansas City, Mo. And finally, each of the women had a different hair color, namely black, brown, gray, red, auburn and blond.

The following pairs played doubles: Howard and Kress against Grace and Susan, McLean and Randolph against Laura and Susan. The two ladies with black and brown hair played first against Howard and McLean, then against Randolph and Kress. The following singles were played: Grace against McLean, Randolph and Lewis, the gray haired lady against

Margaret, Diana and Virginia, the lady from Kansas City against Margaret, Laura and Diana, Margaret against the ladies with auburn and blond hair, the lady from Wichita against Howard and McLean, Kress against Laura and Virginia, the lady from Mt. Vernon against the ladies with red and black hair, McLean against Diana and Virginia, and the girl from Boston against the lady with gray hair.

Finally, Rust played against the lady with black hair, the auburn girl against Diana, Lewis against the girl from Kansas City, the lady from Mt. Vernon against Laura, the one from Kansas City against the auburn one, the woman from Wichita against Virginia, Randolph against the girl from Mt. Vernon, and the lady from Fort Worth against the redhead.

There is only one other fact we ought to know to be able to find the last names, home towns and hair colors of all six wives, and that is the fact that no married couple ever took part in the same game.

21. Six writers in a railroad car.

On their way to Chicago for a conference of authors and journalists, six writers meet in a railroad club car. Three of them sit on one side facing the other three. Each of the six has his specialty. One writes short stories, one is a historian, another one writes humorous books, still another writes novels, the fifth is a playwright and the last a poet. Their names are Blank, Bird, Grelly, George, Pinder and Winch. Each of them has brought one of his books and given it to one of his colleagues, so that each of the six is deep in a book which one of the other five has written.

Blank reads a collection of short stories. Grelly reads the book written by the colleague sitting just opposite him. Bird sits between the author of short stories and the humorist. The short-story writer sits opposite the historian. George reads a play. Bird is the brother-in-law of the novelist. Pinder sits next to the playright. Blank sits in a corner and is not interested in history. George sits opposite the novelist. Pinder reads a humorous book. Winch never reads poems.

These facts are sufficient to find each of the six authors' specialties.

22. Walking around Mt. Antarctic.

There are not many mountains in the world which have not yet been explored and climbed from all sides. But recently an American expedition found some heretofore unknown mountain ranges that look as if they might pose some new problems. Let's call one of the newly discovered mountains Mt. Antarctic and let us imagine that during the next Antarctic expedition a party of explorers decides to walk around the mountain at its base. The complete circuit is 100 miles and, of course, it leads through very cold, inhospitable country. Naturally, in the Antarctic the party has to rely exclusively on the provisions it can take along. The men cannot carry more than two days' rations at a time and each day they have to consume one of these during the march. Each ration is packed as two half-rations. No doubt, it will be necessary to establish depots on the 100-mile circular route, to which the men will have to carry rations from the camp. The greatest distance the party can walk in a day is 20 miles. The questions is, what is the shortest time needed for the party to make the circuit around Mt. Antarctic?

A rather primitive solution would consist of dumping a great number of rations at a point 10 miles from the camp, then carrying half of them to a point 20 miles from the camp, and so on, and so forth, until two daily rations have been dumped at a point 60 miles from camp, from which point the party could reach the camp in two days. With this procedure the party would need about 130 days to complete the circuit. But the number of day's journeys needed for the expedition may be cut considerably by establishing depots around the mountain in both directions from the camp.

However, this solution, too, is by no means the most efficient. The job can be done in a much shorter period, without using any tricks such as fasting or eating the day's ration before starting at dawn to be able to carry two daily rations to the next depot.

23. Horseracing.

Five horses, Condor, Fedor, Tornado, Star and Riotinto, are entered in a race. Their post position numbers—though not in the same order—are 1 to 5. The jockeys riding these horses have the following names: Reynolds, Shipley, Finley, Semler and Scranton. The odds are whole numbers between 1 and 6 inclusive and are different for each horse.

This is the outcome of the race: Reynolds is winner and Finley is last. The favorite, for whom, of course, the lowest odds would have been paid, finishes third. Tornado's position at the finish is higher by one than his post position number. Star's post position number is higher by one than his rank at the finish. Condor's post position number is the same as Fedor's finishing position. Only Riotinto has the same post position and finishing rank. The outsider with the odds 6 to 1 finishes fourth. The horse that comes in second is the only horse whose name starts with the same letter as his jockey's name. The odds on Shipley's horse are equal to his post position number; those on Fedor exceed his rank at the finish by one; those on Condor are equal to his position at the finish and exceed by one the position of Semler's horse at the finish.

What are the names, jockeys' names, post position numbers and odds of these five race horses and in what order do they come in at the finish?

24. Grand Opening.

Some years ago, on the last day of July, a very fashionable Fifth Avenue store opened a new department for expensive dresses. The five sales girls hired for the new department were asked to contribute their share to the show by wearing, every succeeding day for a month after the opening, a different colored dress, as far as their own wardrobes and budgets permitted. As a merchandising experiment the store was kept open every day, Sundays included.

Now this is the way the girls cooperated: None of them owned more than ten dresses, while one had only one good dress which she wore every day. The others rotated their wardrobes so that they always wore their dresses in the same order (for instance, yellow, red, blue, yellow, red, blue, yellow, etc.). None of the girls owned two dresses of the same color and each had a different number of dresses.

One of the store's best customers who dropped in every day all through August and who, of course, didn't miss the slightest detail in the sales girls' wardrobe, kept track of the daily changes and made the following observations: On August 1st, Emily wore a grey dress, Bertha and Celia red ones, Dorothy a green one and Elsa a yellow one. On August 11th, two of the girls were in red, one in lilac, one in grey and one in white. On August 19th, Dorothy had a green dress, Elsa a yellow one and the three other girls red ones. Dorothy had on a yellow dress on August 22nd and a white one on August 23rd. On the last day of the month all five sales girls wore the same dresses as on the first day of August.

These are all the mental notes the faithful customer and observer took. However, when some time later, someone asked her whether she could find out which of the girls had worn a lilac colored dress on August 11th, she was able to answer that question with the aid of a pencil, a piece of paper and her fragmentary recollections. Try whether you can do the same!

CHAPTER III

FADED DOCUMENTS

The "faded documents" dealt with in this chapter are, of course, fictitious. They have never been found in dusty attics or buried treasure chests but have been painstakingly constructed by mathematically minded brains. You will find some of the problems in this chapter very hard to solve and their solution may take you many hours. But you may find comfort in the fact that, in any case, the construction of any such problem is far more difficult and involved than its solution and that skilled mathematicians have sometimes spent weeks devising just one of these "faded documents."

The idea underlying these problems is that somewhere an ancient document has been found, a manuscript so decayed that an arithmetical problem written on the old parchment is for all practical purposes, illegible. So little of its writing is left, in fact, that only a few figures are still recognizable and some faint marks where once other figures had been. And sometimes there is nothing left at all but splotches marking spots where figures have disappeared. But however little is legible, it is sufficient to reconstruct the entire original arithmetical problem.

You need much patience to solve these arithmetical jigsaw puzzles and unless you have a good touch of arithmetical feeling in your finger tips you had better forget about the more complicated problems in this chapter.

Let's start with a simple "faded document" problem and show how this class of puzzles should be tackled. The asterisks, of course, indicate the unrecognizable figures.

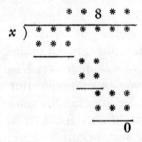

The solution of this puzzle is comparatively simple. The position of the two asterisks in the third line indicates that the divisor x is a two-digit figure, one that divides into the first three digits of the dividend without a remainder. In that case, the fact that both the fourth and fifth digit of the dividend are carried down together would prove that in the quotient, the 8 must be preceded by an 0. It is also clear that the two asterisks in the third line represent a two-digit figure into which another two-digit figure, namely x, divides 8 times. On the other hand, the divisor x divides less than 10 times into a three-digit figure, represented by three asterisks in the fifth row. Only one two-digit figure, namely 12, fulfills both these conditions, and therefore, the divisor x must be 12. x must be bigger than the first two digits of the figure represented by the three asterisks in the third line because otherwise the third digit of that figure would have been carried to the line below. Therefore, the first digit of that figure can only be 1, and that allows us to conclude that the figure is 108, which is divisible by 12 without remainder. Now it is easy to reconstruct the whole division. This is the way it looks:

```
            90809
      12)1089708
         108
          ‾‾‾
           97
           96
          ‾‾‾
          108
          108
          ‾‾‾
            0
```

Like any other branch of entertainment, the "faded documents" have some "classics," old, well-known stand-bys which are particularly accomplished and striking examples of their kind. One such classic is the *Seven sevens,* first published in 1906 in *School World.* To facilitate your solving the other "faded documents" puzzles we shall show how to solve the *Seven sevens.*

(To facilitate the explanation we have framed the problem in a system of coordinated key letters. The 7 in the third line, for instance, is identified by the letters *fC*.)

	a	b	c	d	e	f	g	h	i	j	k	l	m	n	o	p	q	r	s	t	u
A	*	*	7	*	*	*	*	*	*	*	÷	*	*	*	7	*	=	*	*	7	*	*
B	*	*	*	*	*	*	*														
C	*	*	*	*	*	7	*														
D	*	*	*	*	*	*	*														
E			*	7	*	*	*	*													
F			*	7	*	*	*	*													
G			*	*	*	*	*	*	*												
H			*	*	*	*	7	*	*												
I				*	*	*	*	*	*	*											
J				*	*	*	*	*	*												
								0													

The following line of reasoning leads to a solution: Since the 6-digit *F* line is the 7-fold product of the 6-digit divisor, *k* can only be 1. *l* cannot be 0, because if it were, *dF* = 7 would have to be the sum of 0 plus the carried-over ten digit of 7 times m, and this particular digit could not be 7. It would only be 6 ($9 \times 7 = 63$) even if m were 9. On the other hand, l cannot be greater than 2, because if it were, cF ought to be at least 9 and cE greater than 9 (because of the existence of cG), which is impossible.

cG, the difference $cE - cF$, cannot exceed 1, because we have already found that kl cannot be greater than 12, and, the H line being a 7-digit figure, even with $t = 9$, the H line cannot reach 2,000,000. Therefore, cG and cH both must be 1. dG could be 9 or 0. If it were 9, dH must be 9 or 8, because there is no dI. In that case, the H line, or t times the divisor, would begin with 19 or 18. However even if we presume t to be the highest possible numeral, 9, 9 times the divisor, 12 . . . . , couldn't be greater than 11 . . . . . Therefore, both dG and dH are 0.

We haven't decided yet whether l is 1 or 2. If l were 1, m must be 0 or 1, because otherwise dF couldn't be 7. If m were 2, there would be in the F line (the 7-fold product of the divisor) a remainder of 1 (from $7 \times m = 14$) as part of dF, which would make dF an 8 instead of a 7. If the divisor begins with 111, even the factor $t = 8$ could not create a 7-digit H line. Therefore, $l = 2$, and the E line begins with 97, the F line with 87. Since we have ascertained that $l = 2$, $dF = 7$ cannot stem from 7×1, but must originate from 4 (the 4 in $14 = 7 \times 2$) plus 3, which is the remainder from 7 times m. This can only occur if m is at least 4 ($7 \times 4 = 28$) and n greater than 2 (for instance, 3, so that 7×3 would leave a remainder of 2 to be added to 28, leaving a final remainder of 3 as part of dF), or if m is 5 and n no greater than 7. If n were 8, the remainder in question would be 4 ($7 \times 8 = 56$, remainder 5, $7 \times 5 = 35$, plus 5 equals 40, 4 being the remainder to be carried over). Thus, the first 4 digits of the divisor are between 1242 and 1257.

t can be only 8 or 9, because even with the greatest possible divisor, 1257 . ., a multiplication with 7 could not result in a 7-digit H line (divisor times t). On the other hand, assuming t to be 9, a multiplication with 9 of even the lowest possible divisor, 1242 . ., would result in a 1 at dH; we know, however, that $dH = 0$. Hence t must be 8. Moreover, since the divisor must at least begin with 1250 to create a 7-digit H line when multiplied by $t = 8$, we now know that $m = 5$.

Thus, so far, our skeleton has been restored to:

	a	b	c	d	e	f	g	h	i	j		k	l	m	n	o	p		q	r	s	t	u
A	*	*	7	*	*	*	*	*	*		÷	1	2	5	*	7	*	=	*	*	7	8	*
B	*	*	*	*	*	*																	
C	*	*	*	*	*	7	*																
D	*	*	*	*	*	*	*																
E			9	7	*	*	*	*															
F			8	7	*	*	*	*															
G			1	0	*	*	*	*	*														
H			1	0	*	*	7	*	*														
I					*	*	*	*	*	*													
J					*	*	*	*	*	*													
										0													

$gH = 7$ is the sum of (1) the carry-over from t times o ($8 \times 7 = 56 +$ remainder), which may be 5 if 8 times p doesn't reach 40, or otherwise may be 6, and (2) the unit of a multiple of 8, which is always an even number. Since $gH = 7$ is an odd number, the carry-over from $t \times o$ can only be 5 and not 6. Hence we have to consider $gH = 7$ as the sum of 5 plus the unit 2 of a multiple of 8, which can only be 32 or 72. That means that n can only be 4 or 9. Since the first three digits of the divisor are 125 and the fourth digit cannot be greater than 7, n must be 4.

Having now determined 5 of the divisor's 6 digits we can partly develop the F and H lines. The first 3 numbers of the F line are 878, the first 5 numbers of the H line are 10037. Moreover, eE can be only 8 or 9, because there is no remainder from the next to the last digit carried over to $dF = 7$, dG being 0. eG must be 1—and no more than 1—because otherwise there would be no eI. Consequently, $eE = 9$, and eG, eI and eJ are all 1. What we know about the G and H lines indicates that u can only be 1. Thus, we know that both the I and J lines are 12547*, only the last digit still unknown. We now know a great many of the numbers on our "faded document":

	a	b	c	d	e	f	g	h	i	j	k	l	m	n	o	p	q	r	s	t	u
A	*	*	7	*	*	*	*	*	*	*	÷1	2	5	4	7	*	=*	*	7	8	1
B	*	*	*	*	*	*															
C	*	*	*	*	*	7	*														
D	*	*	*	*	*	*	*														
E			9	7	9	*	*	*													
F			8	7	8	*	*	*													
G			1	0	1	*	*	*	*												
H			1	0	0	3	7	*	*												
I					1	2	5	4	7	*											
J					1	2	5	4	7	*											
										0											

hI is 4. *hH*, above *hI*, must be at least 6 (from 8 × 7 = 56). Therefore, *hG* −*hH* must leave a remainder, 1, to be carried over to *gH*. *gI* being 5, a 3 must be above *gH* = 7, which yields a 6 at *fG*.

The *D* line may be 1003... or 1129..., depending on *r* being either 8 or 9. Hence the *C* line, computed by adding the known parts of the *D* and *E* lines, can begin only with 110 or 122. Thus, *cC* is 0 or 2; *cB*, 6 or 7, or 4 or 5, respectively (depending on *cC* being 0 or 2). Only two of these numbers, 6 and 7, occur in that position in the possible multiples of the divisor (1 to 6, because the *B* line has as many digits as the divisor 12547* and 7 is excluded because then *cC* would be 8, like *eF*. These numbers, 6 and 7, at *cB* result from multiplications by 3 and 5, and hence *q* is either 3 or 5. Both 6 and 7 at *cB* result in *r* = 8. (See the beginning of this paragraph.) The *C* line begins with 110, the *D* line with 10037, being identical with the *H* line.

We now have to find out whether *q* is 3 or 5. If *q* is 3, the *B* line would begin with 3764, and *cB*, under *cA* = 7, would be 6. We know that the *D* line begins with 10037 and the *E* line with 979. Since *cC* is 0, a remainder from *dB* must have been carried over and added to the 6 at *cB*. However, this is impossible. We have *dD* = 3 above 7 at *dE*. Thus, *dC* can be only 1 or 0. But then only 4 or 5 could be at *dA*, which would exclude a remainder for *cB*. Therefore, *q* can only be 5.

We have now found the entire quotient and the divisor

with the exception of the unit, p. We determine p by first determining gD. fD can only be 7 or 8, as can easily be determined from what we know about other figures in the f column. fD is the sum of 6 (from 8×7) and a remainder carried over from gD, which can be only 1 or 2. With $r = 8$ as a factor, only 2 or 3 can be chosen for p. Which one is correct, we may decide by trying both, multiplying the divisor by the quotient. We will find $p = 3$. Now we can easily fill in the missing numbers:

	a b	c	d e	f	g h	i	j		k	l	m	n	o	p		q	r	s	t	u
A	7 3	7	5 4	2	8 4	1	3	÷	1	2	5	4	7	3	=	5	8	7	8	1
B	6 2	7	3 6	5																
C	1 1	0	1 7	7	8															
D	1 0	0	3 7	8	4															
E		9	7 9	9	4 4															
F		8	7 8	3	1 1															
G		1	0 1	6	3 3	1														
H		1	0 0	3	7 8	4														
I			1 2	5	4 7	3														
J			1 2	5	4 7	3														
							0													

The classic example of the *Seven sevens* shows how to proceed in reconstructing the more difficult of the "faded documents." If you use this lead, you will have no trouble in solving the following puzzles.

25. The eleven ones.

This division has two peculiarities. First, all ones that occur in the course of the operation are still legible, and second, all whole numbers from 0 to 9, inclusive, occur once in the ten digits that are included in divisor and quotient, together.

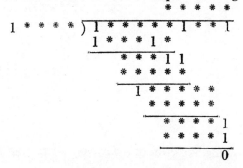

26. Fifteen twos.

In this puzzle, also, all twos which occur in the operation have remained legible, and the divisor and quotient together contain all the one-digit figures from 0 to 9 inclusive.

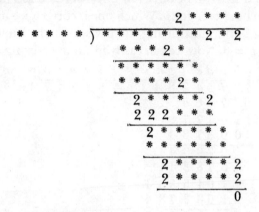

27. The inverted quotient.

In this problem all the threes that occur have remained legible. Moreover, the quotient contains all the numbers in the divisor, but they occur in reverse order.

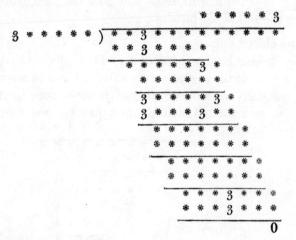

28. Five fours.

The following skeleton of a division represents a comparatively easy "faded document." But, in contrast to the previous problems, not all the fours that occur have been "preserved."

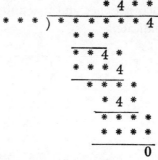

29. Five fours, differently arranged.

Here, all the fours that occur remain readable and, this time, only all the integers from 1 to 9 inclusive occur in the nine figures of the divisor and quotient.

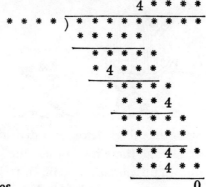

30. Three fives.

The three fives remaining in the following skeleton are sufficient to reconstruct the entire division. However, not every five which occurs in the problem shows up.

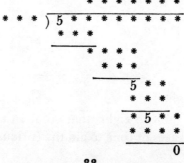

31. Six fives.

The following "faded document" looks simple but its solution requires much patience and ingenuity. Not all the fives that occur are indicated.

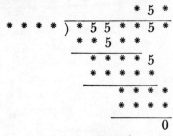

```
                    *  5  *
* * * * ) *  5  5  *  *  5  *
          * *  5  *  *
          ─────────────
            *  *  *  *  5
            *  *  *  *  *
            ──────────────
               *  *  *  *
               *  *  *  *
               ──────────
                        0
```

32. One single seven.

The single seven in this skeleton of a division is the only one that occurs. Though there is no other clue but one single figure, the solution of this puzzle is not difficult.

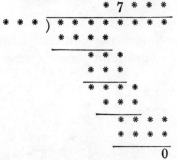

```
            *  7  *  *  *
* * * ) *  *  *  *  *  *  *  *
         *  *  *  *
         ──────────
            *  *  *
            *  *  *
            ───────
               *  *  *  *
               *  *  *
               ──────────
                  *  *  *  *
                  *  *  *  *
                  ──────────
                           0
```

33. Just one eight.

Only one of the two eights that occur in this division has survived the ravages of time to aid the curious.

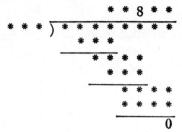

34. Ten eights.

All the eights that occur in this problem are preserved and, as in some previous problems, the divisor and quotient together contain all ten numbers from 0 to 9 inclusive.

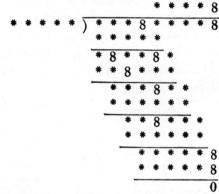

35. Everything faded but the decimal point.

Though this skeleton of numbers shows nothing but one single decimal point it is not too difficult to reconstruct the whole division just with the aid of the one decimal point in the quotient.

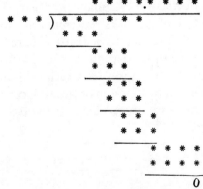

36. Only primes.

Here we have a multiplication for a change. All numerals are prime numbers; moreover, neither ones nor zeroes occur.

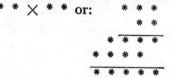

37. Square and cube.

This problem consists of two multiplication skeletons, with no figures whatsoever. The first represents the multiplication of a number by itself, the second the multiplication of the resulting square by the first number so that the final result is the cube of the first number. Here is a hint that will facilitate the solution: The numbers of digits of both products may serve as an indication of the limits within which you will have to look for the first number.

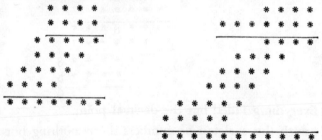

38. Combination of addition and subtraction of squares.

This arrangement of asterisks represents a combination of addition and subtraction of square numbers. The long horizontal middle line of asterisks represents the addition of two square numbers resulting in a third square number. The roots of these three squares are in the proportion of 3 : 4 : 5. The three-decker vertical group of asterisks in the center of the skeleton represents a subtraction of two square numbers resulting in a third square number. There are five different solutions to this problem.

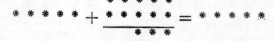

36

CHAPTER IV

CRYPTOGRAMS

Nobody knows how many times the frantic message, *Send More Money,* has been wired. There is nothing peculiar in the combination of these three simple English words, no problem involved—except perhaps for the recipient.

However, on an idle Saturday afternoon, allegedly shortly after World War I, a young employee working in a rural branch of the Western Union happened to write the three words one below the other, time and time again. It was just his way of doodling. Suddenly he realized that such an arrangement of words looked like an addition:

$$
\begin{array}{r}
S\ E\ N\ D \\
M\ O\ R\ E \\
\hline
M\ O\ N\ E\ Y
\end{array}
$$

Having three more hours to stick around and wait for customers who would probably never show up to disturb his leisure on this particular Saturday afternoon, he began replacing each letter in these three words by a number, trying to make the addition come out. He enjoyed this pastime until he indeed obtained a correct addition, as sound and unshakable as two plus two is four. He had invented a new class of puzzle, the numerical cryptogram.

The young fellow in the telegraph office soived his puzzle by trial-and-error, we presume. It is easier and quicker, though in fact it never is really easy, to solve such cryptograms by using logic and all the available weapons of computation. How to set about it we will show by solving the above money problem methodically. All cryptograms should be solved—and constructed, for that matter—in the same way, that is, by proceeding logically and systematically, step by step.

$$\begin{array}{r} SEND \\ MORE \\ \hline MONEY \end{array}$$ being an addition, the letter M can only stand for a 1 (in the third line and consequently, also in the second line), and the letter O must represent the number 0 (also in both the second and third lines). Consequently, the S in the first line can be only 8 or 9. Let's assume that 9 is correct and will help us build up a flawless addition of two four-digit numbers. Being through with the thousands, we tackle the hundreds. We already know that the letter O stands for 0. So we can assume $E + 1$ to equal N. For the tens we may write down the equation $N + R + $ (possibly) $1 = (N - 1) + 10$. Consequently, R should be replaced by either 8 or 9. 9, however, has already been used for S. So we reserve 8 for R. In the $N + R$ equation above we have added 1 only tentatively. Since we had to choose 8 as the equivalent for R, the addition of 1 becomes a must, and that implies that the sum of the units, $D + E$, must be bigger than 10. Because the numbers 8 and 9 have already been used, only two combinations are left, either 6 and 7, or 5 and 7, E cannot be 7, because in that case (look at the tens!), N would become 8, that is, equal to R, which is against the rules of the arts of the cryptogram, which stipulate that each letter stands for a different figure. So we have a choice of 5 or 6 for E. We choose 5, because otherwise N would have to become 7, which is not allowed since 7 is needed for the D. (The last figure, Y, of the sum of D and E must not become a 1 since 1 has already been adopted by M.) In filling the remaining gaps, no mistake is possible, and the puzzle is solved:

$$9\ 5\ 6\ 7$$
$$1\ 0\ 8\ 5$$
$$\overline{1\ 0\ 6\ 5\ 2}$$

Since the *Send More Money* problem first turned up, many cryptograms have been devised. Often, the word combinations make sense, more often, they do not, but in any case, when each letter has been identified with one—and only one—digit, and vice versa, the resulting computation should be correct.

The above cryptogram can have only one solution. Others are not always unequivocal and sometimes may have a great number of solutions, so that in many cases the question may be asked, How many solutions are possible?

Two variations of the above cryptogram follow, both of which permit of several solutions.

Wherever more than one solution is possible, no more than two will be given—the one with the smallest sum and the one with the largest. We leave it to you to find all the possible solutions in between.

39.

$$S\ E\ N\ D$$
$$M\ O\ R\ E$$
$$G\ O\ L\ D$$
$$\overline{M\ O\ N\ E\ Y}$$

32 solutions are possible.

40.

$$S\ A\ V\ E$$
$$M\ O\ R\ E$$
$$\overline{M\ O\ N\ E\ Y}$$

4 solutions are possible.

41.

$$J\ U\ N\ E$$
$$J\ U\ L\ Y$$
$$\overline{A\ P\ R\ I\ L}$$

42.

$$AB + BA + B = AAB$$

43. Two equations.

$$AB \times C = DE$$
$$FG - DE = HI$$

Only two solutions are possible.

44. Nine A's and three B's.

$$AAA \times AAA + AAA = AAABBB$$

45. A two-step multiplication.

$$A \times B = CD,$$
$$CD \times EF = GHI.$$

All nine digits from 1 to 9 inclusive are represented. Only one solution is possible.

The series of cryptograms which follow are a little more involved. They combine three arithmetical operations—multiplication, addition and subtraction. Otherwise, the rules are the same: Only one numeral for each letter; only one letter for each numeral. The two multiplications are related in the following way: The sum of the two first factors plus the sum of the two second factors equals the difference between the two products.

46.

$$
\begin{array}{ccccc}
ab & \times & ac & = & acb \\
+ & & + & & - \\
ad & \times & e & = & aaf \\
\hline
fd & + & fd & = & de
\end{array}
$$

47.

$$
\begin{array}{ccccc}
ab & \times & cd & = & cdb \\
+ & & + & & - \\
ae & \times & ae & = & afd \\
\hline
ce & + & eb & = & de
\end{array}
$$

48.

$$
\begin{array}{ccccc}
a & \times & ba & = & cda \\
+ & & + & & - \\
eb & \times & eb & = & efa \\
\hline
eg & + & ag & = & hg
\end{array}
$$

49.

$$
\begin{array}{ccccc}
a & \times & abc & = & dec \\
+ & & + & & - \\
fg & \times & fg & = & adg \\
\hline
fe & + & hg & = & hai
\end{array}
$$

50.

$$a \times bc = dba$$
$$+ \quad + \quad -$$
$$ce \times ce = fac$$
$$\overline{gh + cii = cgh}$$

51.

$$ab \times cdb = efgb$$
$$+ \quad + \quad -$$
$$fa \times fa = eafb$$
$$\overline{cdi + dii = hdi}$$

52.

$$ab \times cdbab = ekbaib$$
$$+ \quad + \quad -$$
$$bae \times bae = fghfib$$
$$\overline{bdk + cbadk = cbhkk}$$

53.

$$ab \times cdefaab = ghddeiecb$$
$$+ \quad + \quad -$$
$$gkbcd \times gkbcd = gebhceacb$$
$$\overline{gkcff + cdhkhff = cifcbff}$$

41

CHAPTER V

How Old Are Mary and Ann?

All problems in this chapter are similar. You are requested to find out how old a number of people are whose ages now or in the past or in the future have certain arithmetical relations to one another. The difficulty with such problems is to find the equation that covers all the given relations, which, incidentally, may be highly involved. Often, this task is much more difficult than it looks at first glance and you may get into a labyrinth from which escape seems rather hopeless.

Take Mary and Ann, for instance. Some years ago, many people cudgeled their brains about the ages of these two girls, without getting anywhere. The problem looks rather harmless: The present sum of the ages of the two girls is 44. Mary is twice as old as Ann was when Mary was half as old as Ann will be when she is three times as old as Mary was when she was three times as old as Ann.

It is pretty hopeless to tackle this problem simply by reasoning through all the given relations beginning with the first statement. Very soon the thread leading out of the labyrinth is lost altogether. It is easier to attack this puzzle with simple arithmetical tools and advance step by step, for instance as follows:

	Ann's age	Mary's age	Age difference
When Ann was............	x		
Mary was three times as old..		$3x$	$2x$
If Ann were three times as old as Mary was at that time, namely	$9x$		
and if Mary were half as old, that is		$4\frac{1}{2}x$	

42

Ann still was $2x$ years younger,
namely . . . . $2\frac{1}{2}x$
Today, Mary is twice $2\frac{1}{2}x$, that
is . . . . $5x$
and Ann still is $2x$ years
younger, that is. . . . . $3x$
Today, Ann and Mary together
are 44, which yields the
equation . . . . $3x + 5x = 44$

Consequently, x equals $5\frac{1}{2}$; Ann, being $3x$, is $16\frac{1}{2}$; Mary, being $5x$, is $27\frac{1}{2}$. You see, it wasn't quite so simple to find the equation which yields the age of these two girls. Most of the following questions will be much easier to answer.

54. Mr. MacDonald's prolific family.

When MacDonald was over 50 but less than 80 he told a friend: "Each of my sons has as many sons as brothers, and the

number of my years is exactly that of the number of both my sons and grandsons." How many sons and grandsons had McDonald?

55. The age of Mr. and Mrs. Wright.

Says Mrs. Wright: "Today, I and my two children together are as old as my husband. What will be the ratio of my husband's age to my present age when my children together will be as old as I?

56. Foreman and manager.

The foreman in a little factory is 48. How old is the plant manager if the foreman is now twice as old as the manager was when the foreman was as old as the manager is now?

57. Four generations.

There is an old Civil War veteran who has a son, a grandson, and a great-grandson. He and his great-grandson together are as old as his son and grandson together. If you turn his age around (that is, take the units for tens and vice versa) you will get the age of his son, and if you turn around the age of his grandson, you will get the age of his great-grandson. To make things a little easier for you, all four numbers are prime numbers. This will suffice for you to find the age of these four people.

58. The silver wedding anniversary.

Somebody asked Mrs. Reynolds how long she had been married. Her answer was: "If I were married twice as long as I am it would still be half again as long as I am now married until I could celebrate my silver wedding anniversary." How long has Mrs. Reynolds been married?

59. Three years make quite a difference.

My age is thrice my age three years hence minus thrice my age three years ago.

60. One hundred years.

Today, Jim Higgins and his three sons are together a hundred years old. Moreover, Jim is twice as old as his oldest son who, in turn, is twice as old as his second son who, in turn, is twice as old as the youngest. How old are the members of the Higgins family?

61. Mrs. Hughes and her family.

Today, Mrs. Hughes is three times as old as her son and nine times as old as her daughter. At 36, Mrs. Hughes will be twice as old as her son and three times as old as her daughter. When Mrs. Hughes celebrates her silver wedding anniversary her children together will be just as old as she. How old is she today, how old are her children and how old was she when she married?

62. Involved age relations.

Bill is twice as old as Sam was when Bill was as old as Sam is. When Sam is as old as Bill is now, both of them together will be 90. How old are they?

63. June and December.

Eighteen years ago, the Wellers got married. At that time, Mr. Weller was three times as old as his wife. Today, he is only twice as old as she is. How old was Mrs. Weller when she got married?

64. Two friends.

"Do you know," Marion asked Betty, "that in exactly seven years the sum of our ages will be 63?" "That's funny," answered Betty, "I have just discovered something else. When you were as old as I am now you were just twice as old as I was then." How old were these two friends?

65. A seemingly simple problem.

Two brothers together are 11. One is 10 years older than the other. Find out how old both of them are (and don't make the same mistake you almost made when you figured the price of that bottle in problem No. 1).

66. How old are the four children?

The sum of the ages of the four children of a family is equal to one half their father's age. The difference of the squares of the ages of the oldest and youngest child equals the father's age which is also equal to twice the sum of the differences of the squares of the ages of the two younger and the two older children. In 16 years, the sum of the ages of the children will exceed the father's age by his present age. How old are the father and the four children?

CHAPTER VI

WOLF, GOAT AND CABBAGE — AND OTHER ODD COINCIDENCES

This chapter deals with difficult river passages. We will offer you only a few of the many existing puzzles of this class, which is a staple of arithmetical entertainment and thus warrants treatment in a chapter of its own. Some river-crossing problems can boast of venerable age, which may serve as an excuse for the odd combination of people or objects which have to be moved over the imaginary river. These, moreover, have to be moved in a rowboat, as these problems were devised at a time when there were hardly any other means of crossing rivers. (Swimming is against the rules governing these puzzles.)

These little problems have hardly anything to do with figuring, though they do have something to do with figures. No equations can be written down. You will just have to think hard and then try whether your method is the one that gets your cargo across most quickly. One of these ancient problems is the one of the wolf, the goat and the cabbage. Alcuin, friend and teacher of Charlemagne, is supposed to have invented it and to have inserted it into the lessons he gave his imperial ward (though we do not know whether it contributed much to the legendary wisdom of the great Frankish King and Roman Emperor). However, since then it seems to have found its place in ancient arithmetic books as well as in modern puzzle collections.

In case you never heard of it, this is the problem—though we do not pretend that it could arise in, say, contemporary Minneapolis. Someone owns, of all things, a wolf, a goat and a cabbage. Coming to a river, he charters a rowboat which is so small that in addition to himself it can only carry either the wolf, the goat or the cabbage. If left alone, the wolf would eat the goat, and the goat—if not yet devoured by the wolf—

would eat the cabbage. How can one avoid such a catastrophe while ferrying the precious cargo across the river?

Well, this problem is easy to solve. First, the owner of this odd assortment of fauna and flora has to get the goat across, since the wolf dislikes cabbage. Then he rows back and fetches the cabbage. On the return trip he takes his goat along in order to keep the cabbage intact. He now leaves the goat alone and takes the wolf along to join the cabbage. Then he rows back for the last time to get his goat across, which solves his problem without mishap.

Another order of events is possible: goat across, wolf across, goat back, cabbage across, goat across.

In the XVIth century, the Italian mathematician, Tartalea, devised the following more involved variant of the river-crossing problem: A party of three jealous husbands and their wives who want to cross a river have only one boat, and this boat accommodates but two people. As a matter of precaution they agree that at no time, neither on this nor the opposite shore, nor in the boat, shall any of the wives be in the company of one or two men unless her husband is present to watch her. It is not easy to manage the passage but it can be done if this order of events is followed:

First, all three wives have to get to the other shore, which can be attained by letting, first, two of them row across and then one of these come back for the third. Now one of the three wives rows back and, for the time being, remains with her husband on the near shore. The other two men, meanwhile, join their wives. Then one couple returns to the near shore. This now leaves one couple on the other shore while two couples are on this side. Next, the men of these two couples cross the river, leaving their wives on the near side. Soon, the two women are joined by the returning third wife. Then, two of the wives row across and one of them finally comes back to fetch the third. Altogether that makes six crossings and five return trips, which is the minimum of crossings required for the task.

If there are more than three couples with such inconveniently jealous husbands the problem can only be solved if the capacity of the boat is simultaneously raised to $n - 1$, n being the number of couples. Amazingly, for four couples using a boat holding three people, only five crossings and four return trips are needed. But it is impossible to get four couples across in a two-seater boat unless one finds a way out by allowing for an island in the middle of the river which may serve as a transfer stop. This variant was first described in 1879 by E. Lucas in his *Recreations*. It is not easy to solve and since you may want to test your ingenuity we will relate the problem as follows:

67. Four jealous husbands and one island.

Four couples have to cross a river with the aid of one boat with a capacity of only two people. Nowhere, neither on land nor in the boat, is any of the women supposed to be left in the company of any of the men unless her husband is present. How

can the transfer be executed if there is an island in the middle of the river where the boat may land and leave some of its cargo? By the way, it is good to use playing cards as a visual aid for solving this puzzle.

68. Getting the nuggets across.

During the Alaska gold rush, three prospectors who had struck it rich had to cross the Yukon. But they could find only a small row boat with a capacity of either two men or one man and a bag of gold nuggets. Each of them owned a bag of the precious metal but the contents were not the same. Smith's gold was worth $8000, Jones' $5000 and Brown's $3000. None of them trusted either of the other two, and so, after some argument, they agreed that the passage should be arranged in such a way that none of them, either ashore or in the boat, should be in the presence of nuggets worth more than those he owned. How did the men have to proceed, to get themselves and their property across the river in no more than 13 crossings, all told? (Returns across the river count as crossings.)

69. Missionaries and cannibals.

Three missionaries and three cannibals find themselves in a similar mess the prospectors were in, that is, in possession of a two-seater row boat with which to cross a river. Since the man-eating savages are ravenously hungry it is imperative

that on neither shore are they ever in the majority. The priests' difficulties are aggravated by the circumstance that of the three anthropophagous barbarians only one knows how to row, though all three missionaries are masters of that art. How does this uneasy company of six get across the river?

CHAPTER VII

CLOCK PUZZLES

On our wrists or in our vest pockets we carry veritable treasure-boxes of arithmetical problems. The two hands of the watch which move around with uniform, though different, speeds, always and forever passing each other, are a constant challenge to the mathematical-minded. Millions of curious souls have probably figured out how often the two hands pass each other in a 24-hour day. But this is one of the simplest problems our watches or grandfather clocks can put to us. Many others are pretty involved and if we try to solve them we may be led into the realm of indefinite or Diophantine equations. (One equation with two unknowns or two equations with three unknowns, etc.) This is true particularly if we introduce a third racing factor, the second-hand, for then we may get problems which require some mathematical background and experience and a great deal of constructive ingenuity. On the other hand, most clock puzzles offer one distinct advantage. In most cases you may find, or at least check, the answers just by moving the hands of your wrist watch. So if you want to solve involved clock problems fairly it might be a good idea to remove your wrist watch and lock it in the drawer before tackling the following puzzles.

70. Race of the hands.

How often and when in their eternal race will the hour-hand and the minute-hand meet? This question is not difficult to answer and here is another almost as simple: how often will the two hands be at right angles to each other?

71. Two minutes apart.

The long hand of a very accurate time piece points exactly at a full minute while the short hand is exactly two minutes away. What time is it?

72. Exchanging the hands.

At what time can the position of the long and short hands be reversed so that the time piece shows a correct time? How many such exchanges of the hands are possible within 12 hours?

At 6:00, for instance, such an exchange is impossible, because then the hour-hand would be directly on 12 and the minute-hand on 6, that is, 30 minutes after the full hour, which doesn't make sense.

73. Three hands.

Imagine a railroad station clock with the second-hand on the same axis as the two other hands. How often in a 24-hour day, will the second-hand be parallel to either of the two other hands?

74. The equilateral triangle.

Let us imagine that all three hands of a clock are of the same length. Now, is it possible that at any time the points of the hands form an equilateral triangle? In case this should prove impossible, at what moment will the three hands come closest to the desired position?

75. Three clocks.

At noon, April 1st, 1898, three clocks showed correct time.

One of them went on being infallible while the second lost one minute every 24 hours and the third gained one minute each day. On what date would all three clocks again show correct noon time?

76. The two keyholes.

There are two keyholes below the center of the face of a clock, arranged symmetrically on either side of the vertical center (12 - 6) line. The long hand moves over one of them in three minutes, between the 22nd and 25th minutes of every hour, and over the other one also in three minutes, between the 35th and 38th minute. The short hand is as wide as the long hand and long enough also to move over the two keyholes. For how long during 12 hours will neither keyhole be covered by either of the hands?

77. The church clock.

A church clock gains half a minute during the daylight hours and loses one-third of a minute during the night. At dawn, May first, it has been set right. When will it be five minutes fast?

78. Reflection in a mirror.

Imagine that we place a clock, the face of which has no numerals but only dots for the hours, before a mirror and observe its reflection. At what moments will the image show correct time readings? This will happen, for example, every hour on the hour, but this represents only some of the possible solutions.

CHAPTER VIII

TROUBLE RESULTING FROM THE LAST WILL AND TESTAMENT

This chapter deals with a number of problems concerning estate-sharing difficulties resulting from curious wills. One such problem is supposed to have arisen in ancient Rome. It was caused by an involved testament drawn up by the testator in favor of his wife and his posthumously born child and it became a legal problem because the posthumous child turned out to be twins.

The will stipulated that in case the child was a boy the mother was to inherit four parts of the fortune and the child five. If the child was a girl she was to get two parts of the fortune and the mother three. But the twins born after the testator's death were a boy and a girl. How was the estate, amounting to 90,000 denarii, to be distributed?

An exact mathematical solution of this problem is impossible. Only a legal interpretation can lead to a solution which abides best by the spirit, if not the letter of the testament. It may be assumed that the testator was primarily concerned with the financial security of his wife. Therefore, at least four ninths of the fortune—assuming the least favorable of the two alternatives—should fall to her, while the remainder should be distributed between the two children in the proportion of 25 to 18, which is in accordance with the $\frac{5}{9}$ to $\frac{2}{5}$ proportion set forth in the two parts of the will. Therefore, the mother should inherit 40,000, the boy 29,070 and the girl 20,930 denarii.

Or we may assume that the testator under all circumstances wished to let his son have five ninths of his estate. In that case, the remainder would have to be distributed proportionately between the wife and the daughter.

However, probably the most equitable partition would be first to divide the fortune into two equal parts and to let mother and son participate in one part, mother and daughter in the other, all three to be given their shares in accordance with both alternatives provided for in the will. In that case, the mother of the twins would receive 20,000 + 27,000 = 47,000 denarii, and the son and daughter would get 25,000 and 18,000 denarii, respectively.

Quite a different problem, that of the division of the camel herd, is also of venerable age. A Bedouin left his camel herd to his three sons to be distributed in such a way that one gets one eighth, the second, one third and the third, one half of it.

When the Bedouin died the herd consisted of 23 camels. Soon the sons found out that it was impossible to distribute the animals in accordance with the father's will. Finally, a friend came to their rescue by offering to lend them, just for the distribution, one of his camels. Now the herd consisted of 24 camels and each of the three sons could get his share, one three, the second eight and the third twelve camels. And moreover, the camel which the helpful friend had lent them for the execution of the old man's will was left over and could be returned to its rightful owner. Now, how can this contradiction be explained?

The first question, of course, is whether the testator has really disposed of his entire estate. He has not, in fact, because $\frac{1}{8} + \frac{1}{3} + \frac{1}{2}$ of anything, whether a herd of camels or the sum of one-thousand dollars, is never the whole thing but only $\frac{23}{24}$ of it. Therefore, even if the sons, to execute their father's last will faithfully, had butchered all 23 camels and distributed the meat exactly in accordance with the stipulation of the testament, a remainder amounting to a little less than the average weight of one camel would have been left over. If the herd is augmented by one animal, however, $\frac{23}{24}$ of the herd can be distributed among the sons and the remainder, that is, $\frac{1}{24}$, can be restored to the friend, because $\frac{1}{24}$ of the herd now amounts to exactly one camel.

Yes, certainly, this procedure is a detour, not in accordance with the letter of the will and legally faulty—which proves that an inaccurate last will is just no good and may cause loads of trouble.

79. Two similar last wills which turned out to be quite different.

A testator drew up a will which stipulated that his fortune should be distributed among his children in the following way: The eldest should get $1,000 and one third of the remaining estate, the second should receive $2,000 and again one third of the remainder, and so on down to the youngest child. At the last no remainder should be left for the youngest child, only the outright sum. It is obvious that he had more than two children, but just how many is not revealed. After a number of years, when his fortune had considerably increased and, moreover, he had acquired two more children, he drew up a new will with exactly the same stipulations. What was the size of the estate in both cases and how many children had the testator? The fortune consisted of a full amount of dollars and no cents.

80. The wine dealer's testament.

A wine dealer left to his three sons 24 wine casks, five of which were full, eleven only half full and eight empty. His last will stipulated that each of the sons should inherit the same quantity of wine and the same number of casks. Moreover, since each of the casks contained wine of a different vintage, mixing and decanting was out of the question. How could the wine dealer's last will be fulfilled?

The solution of this problem may be found by trial and error, though solving the problem with the aid of equations is quite simple. There are three different solutions.

81. The bowling club's heritage.

There was once an old New York bowling club with twenty members. One of the members died and it turned out that he had left to the remaining 19 members of his club all that was to be found in his wine cellar. The will stipulated that each of his friends should get the same quantity of wine and the same number of kegs. Decanting was prohibited according to the will. When the stock was taken, only three-gallon kegs were found in the wine cellar. 83 of them were full, 83 were two-thirds full, 38 were one-third full and 23 were empty. The 19 members of the bowling club faithfully carried out their late friend's last will. It turned out that each of the 19 men had to receive a different combination of the full, partly full and empty kegs. How were the contents of the wine cellar distributed? (You can hardly solve this puzzle without using equations.)

82. The missing dress.

The day after the owner of a fashionable Los Angeles store had bought an assortment of dresses for $1800, he suddenly died. Right after the funeral, the merchant's sons took stock and found 100 new dresses, but no bills. They didn't know how many dresses their father had acquired, but they were sure that there had been more than 100 and that some had been stolen. They knew how much he had spent for the dresses and they remembered having heard him remark casually that if he had closed the deal one day earlier he would have obtained 30 dresses more for his money and each dress would have been $3 less. This casual remark enabled the sons to find out how many dresses were stolen. Can you figure out how many dresses the old man had bought?

83. Six beneficiaries of a will.

Three married couples inherit altogether $10,000. The wives together receive $3,960, distributed in such a way that Jean receives $100 more than Kate, and May $100 more than Jean. Albert Smith gets fifty per cent more than his wife, Henry March gets as much as his wife, while Thomas Hughes inherits twice as much as his wife. What are the three girls' last names?

CHAPTER IX

SPEED PUZZLES

Speed is the essence and perhaps the curse of our age. The majority of all the people we call civilized are always in a hurry, always running their lives according to a rigid time schedule. They are lost when their watches are out of order, for they have to rely on the time tables of airplanes and railroads, or the schedules of street vehicles or on the speed of their own cars. They continually glance at clocks and watches and, unconsciously, carry out simple computations involving speed, time and distance.

This chapter, too, is only concerned with simple computations, although it occasionally involves more complex logical operations. All the puzzles which follow have to do with only the simplest form of motion—uniform motion, that is, motion with constant speed. From high school physics you will remember that this type of motion is controlled by the formula $d = s \times t$, that is, the distance traveled equals the speed multiplied by the time elapsed. You won't need any more complicated formula for the following problems.

84. Encounter on a bridge.

On a foggy night, a passenger car and a truck meet on a bridge which is so narrow that the two vehicles can neither pass nor turn. The car has proceeded twice as far onto the bridge as the truck. But the truck has required twice as much time as the car to reach this point. Each vehicle has only half its forward speed when it is run in reverse. Now, which of the two vehicles should back up to allow both of them to get over the bridge in minimum time?

85. How long is the tree trunk?

Two hikers on a country road meet a timber truck carrying a long tree trunk. They speculate on how long the trunk is but their guesses do not agree. One of them is rather mathe-

matical-minded and quickly devises a simple method for figuring the length of the tree. He knows that his average stride
is about one yard. So he first walks along the tree trunk in
the same direction as the slowly moving vehicle, and counts
140 paces. Then he turns back and walks in the other direction. This time it takes him only 20 paces to pass the tree
trunk. With the aid of these two measurements he is able
to figure the length of the trunk. Are you?

86. Streetcar time table.

Another pedestrian walks along the double tracks of an
urban streetcar. The cars in both directions run at equal
intervals from each other. Every three minutes he meets a car
while every six minutes a car passes him. What is the interval
at which they leave from the terminal?

87. The escalator.

There are very long escalators in some New York subway
stations. You don't have to climb them, since the moving steps
will do the job for you. However, two brothers have to get to
a baseball game and are in a hurry and so they run up the
moving steps, adding their speed to that of the escalator. The
taller boy climbs three times as quickly as his little brother,
and while he runs up he counts 75 steps. The little one counts
only 50. How many steps has the visible part of this New
York escalator?

88. Two boys with only one bicycle.

Two boys want to see a football game together and want
to get to it as quickly as possible. They have only one bicycle
between them and decide to speed up their arrival by having
one of them use the bike part of the way, leave it at a certain
point and walk the rest of the route, while the other is to
mount the bicycle when he reaches it and bike to the common
point of destination, which the two are to reach simultaneously. This problem and its solution are, of course, elementary. The question may be varied by stipulating several
even numbered stages and solving by dividing the route into
an even number of equal stages and dropping and picking up

the bike at every stage. In any case, each boy walks half the way and rides the other half.

This variant, however, is less elementary: The route leads over a hill mounting and sloping equally, and the summit of the hill is exactly halfway along the route. The bicyclist can climb the hill with twice the speed of the pedestrian, while downhill he attains three times the pedestrian's speed. Walking speed is the same uphill as down. At what point on the route does the boy who uses the bicycle first have to drop the vehicle so that both arrive simultaneously and most quickly?

89. A motorcycle just in time.

Two men want to catch a train at a rural railroad station 9 miles away. They do not own a car and they cannot walk faster than 3 miles an hour. Fortunately, just as they start on their journey a friend of theirs comes along on a motorcycle. They tell him about their train leaving in $2\frac{1}{2}$ hours while they are afraid it will take them 3 hours to reach the station. The friend offers to help them out. He will take one of them along for part of the journey. Then, while this man continues his trip on foot, the motorcyclist will return until he meets the other man who, meanwhile, will have started his journey on foot. Then the second man will go along as passenger all the way to the station. The motorcyclist's speed is 18 miles an hour. Where should he stop to let the first man dismount and to return for the second man if he wants both men to reach their goal simultaneously? How long will the whole journey take?

90. Bicycling against the wind.

If it takes a bicyclist four minutes to ride a mile against the wind but only three to return with the wind at his back, how long will it take him to ride a mile on a calm day? Three and a half minutes, you probably figure. Sorry, you are wrong.

91. Two railroad trains.

A well known member of this particular family of puzzles is the one about the two railroad trains: How long will it take two trains, lengths and speeds of which are known, to pass each other when meeting en route? This problem is simple and anybody who has solved the one about the tree trunk (No. 85) will know how to tackle it. So we shall omit this one in favor of another which also concerns two trains passing each other.

At 8:30, a train leaves New York for Baltimore; at 9:15, another train leaves Baltimore for New York. They meet at 10:35 and arrive at their destinations simultaneously. When will that be, if both trains have moved with uniform speed and there were no stops or other interruptions on either trip?

92. Two more trains.

Two trains start simultaneously, in opposite directions, from the terminals of a two-track road. One train arrives at its destination five hours earlier than the other, and four hours after having met the other train. What is the ratio of the speeds of the trains?

93. Accident on the road.

One hour after a train has left Aville for Beetown an accident occurs that compels the engineer the proceed with only three-fifths of the time table speed. The train reaches Beetown two hours late. An angry passenger wants to know why the train is so late, and the engineer informs him that the train would have arrived 40 minutes earlier if the location of the accident had been 50 miles nearer to the destination. This information did not pacify the angry traveler entirely but it may help you find out how far Beetown is from Aville; how long the trip should have taken according to the time table, and what the train's usual speed is.

94. A complicated way to explore a desert.

Nine explorers, with as many cars, go about exploring a desert by proceeding due west from its eastern edge. Each car can travel 40 miles on the one gallon of gas its tank holds, and can also carry a maximum of nine extra gallon cans of gas which may be transferred unopened from one car to another. No gas depots may be established in the desert and

61

all the cars must be able to return to the eastern edge camp. What is the greatest distance one of the cars can penetrate into the desert?

95. Relay race.

At a relay race the first runner of the team hands on the baton after having raced half the distance plus half a mile, and the second after having run one third of the remaining

distance plus one third of a mile. The third reaches the goal after having raced one fourth of the remaining distance plus one fourth of a mile. How long was the course?

96. Climbing a hill.

Climbing a hill, Jim makes one and a half miles an hour. Coming down the same trail he makes four and a half miles an hour. The whole trip took him six hours and he did not even stop to enjoy the view from the mountain top. How many miles is it to the top of the hill?

97. Journey around the lake.

Between three towns, Ashton, Beale and Caster, all situated on the shore of a huge lake, there is regular steamship communication. Ashton and Beale, both on the south shore, are 20 miles apart. Every day two steamships leave simultaneously from Ashton, the westernmost of the two towns, to make their several daily trips around the lake townships. They each have a uniform speed but the two speeds are different. One makes the trip by way of Caster, in the north. The other boat travels around the lake in the other direction, by way of Beale in the east. They meet first between Beale and Caster, 5 miles from Beale, then again exactly in the middle between Ashton and Caster, and for the third time exactly between Ashton and Beale. The boats travel in straight lines; the time they lose at the stops is negligible. How far are the distances between the three towns and what is the ratio of the speeds at which the two boats travel?

98. Xerxes' dispatch rider.

When Xerxes marched on Greece his army dragged out for 50 miles. A dispatch rider had to ride from its rear to its head, deliver a message and return without a minute's delay. While he made his journey, the army advanced 50 miles. How long was the dispatch rider's trip?

99. Two friends and one dog.

Two friends, Al and Bob, and their dog, spent their vacation in the Maine woods. One day, Al went on a walk, alone, while Bob followed him an hour later, accompanied by the dog. He ordered the dog to follow Al's trail. When the dog reached Al, Al sent him back to Bob, and so on. The dog ran to and fro between the two friends until Bob caught up with Al, who happened to be a slow walker. Indeed, Al was making no more than $1\frac{1}{2}$ miles an hour, while Bob made 3. The dog's speed was 6 miles an hour. Now, what is the distance the dog ran to and fro until Bob caught up with Al? We may presume that the dog lost no time playing with his two masters or hunting rabbits.

100. Short circuit.

When a short circuit occurred, Ethel searched her drawer for candles and found two of equal length. One was a little thicker than the other, and would burn five hours, the thinner one would burn only four. However, the electrical trouble did not last as long as the life expectancy of the candles. When it was all over, Ethel found that the stump of one of the candles was four times as long as that of the other. Since Ethel was mathematically minded she had no trouble figuring out how long it had been until the blown-out fuse was replaced. Well, how long had the current been cut off?

101. Three candles.

Three candles are held in the same plane by a three-branched candlestick so that the bases of all three are at the same height above the table. The two outer candles are the same distance away from the one in the center. The thin center candle is twice as long as each of the two others and its burning time is four hours. The two outer candles are of different thicknesses from the center candle and from each other. One can burn $4\frac{1}{2}$ hours and the other one 9.

At the beginning of a Thankgiving dinner the candlestick was placed on the table and the three candles were lighted. When the dinner was over the candles were snuffed out by the host. Then one of the guests did a little measuring and found that the heads of the three candles formed a slanting straight line. How long had the feast lasted?

CHAPTER X

Railroad Shunting Problems

Life at a railroad switching yard is full of change and interest although the layman seldom understands the distribution problems which are often solved by very complicated movements of engines and cars. He marvels at the skillful maneuvering by which cars seemingly hopelessly blocked are freed. He is fascinated to see scores of cars from various tracks assembled into a long freight train. New problems turn up continually, and to solve them with the least loss of time demands great experience and ingenuity.

There are turntables at the switching yards which greatly facilitate the job of shifting freight cars and locomotives. However, both the shunting problems that occur in the yard and those that must be solved anywhere on the line are intricate and interesting enough to warrant inclusion in a mathematical puzzle collection.

It is worth noting that playing cards are an excellent visual help for solving most railroad puzzles, such as the following:

102. The short siding.

On a single-track railroad going east and west two trains, E and W, meet as shown in Fig. 1. With the aid of a dead-end

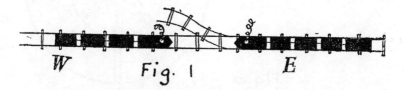

W   Fig. 1   E

siding branching off at the point where the two locomotives stop, the trains manage to continue their runs. Train E has four cars, train W three. The siding, which can be entered only from the east, can receive only one car or one locomotive at a time. How can the two trains maneuver to pass each other with a minimum of moves?

103. Passing on the main track.

An express train *A* has to pass a local *B* with the aid of a dead-end siding, as shown in Fig. 2. There is room for ten

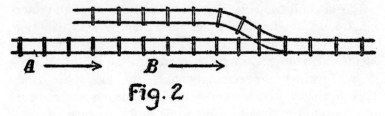

Fig. 2

cars on the siding while both trains have 15 cars each. How will the engineers proceed?

104. A complicated turntable problem.

A train consists of the locomotive *L* and 7 cars, *A*, *B*, *C*, *D*, *E*, *F* and *G*. With the aid of a turntable, the train is supposed to be turned around so that the order of the engine and cars remains unchanged. The turntable consists of 8 radial tracks, each of which can receive several cars. As shown in Fig 3, the

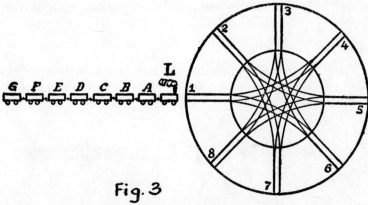

Fig. 3

tracks on the turntable are partly connected. Thus, for instance, a passage from track 1 to tracks 4 and 6 is possible. How can the train be turned around with the fewest possible moves?

105. Keeping out of a trap.

From the main track, two shunting tracks, I and II, branch off, both leading to a short track, *a*, which is just long enough to receive one car, though not an engine (Fig 4.).

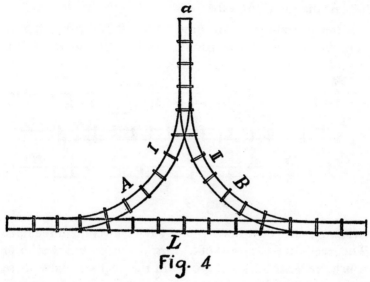

Fig. 4

Car *A* stands on I, Car *B* on II, while the donkey engine, *L,* stands ready on *M*. Now, how can *L* shunt *A* on to II and *B* on to I without cutting itself off from access to the main track?

106. Siding with turntable.

Both ends of a short siding merge with the main track *m - n*. Between the two sections, I and II, of the siding there is a turntable which can carry only one car but no engine. On

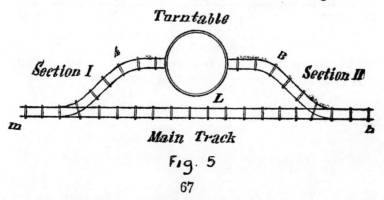

Fig. 5

each section of the siding stands a car (*A* and *B*) as indicated in Fig. 5. The locomotive *L* is on the main track. The problem is to move *A* onto II and *B* onto I while the locomotive has to be on *m* - *n* after it has finished the job.

107. At the switching yard.

A freight train runs down the track *AB* in the direction *A* (Fig. 6). There are two parallel tracks, *CD* and *ES*. *S* is a

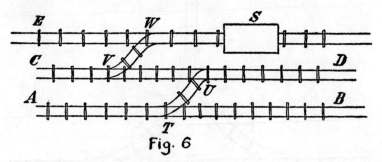

Fig. 6

switching yard. Switches *TU* and *VW* connect the three rails. The freight train consists of 21 cars, the ninth and twelfth of which are destined for *S*. How will the engineer move these two cars to *S* with a minimum of moves?

108. Shunting problem at a station.

At a railroad station a single-track line, *a-b*, has a siding, *d*, as shown in Fig. 7. On each of the two tracks there is space

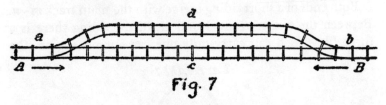

Fig. 7

for 20 cars. At this station, two trains, *A* and *B*, meet with 24 cars each. How can they pass each other?

CHAPTER XI

AGRICULTURAL PROBLEMS

Most of the following problems have little practical significance for agriculture in the age of tractors, combines, diesels, harvesters and marketing cooperatives. Some of them, however, did have significance at a time when the farmer carried his twenty eggs or two geese to the market place fifteen miles away, sold them for what we nowadays have to pay for a pair of frankfurters—and still made a living.

At that time, even famous mathematicians condescended to solve farm problems which were too involved for illiterate peasants to tackle. Some of these old problems may still stump a mathematically well-equipped amateur. Take, for instance, the following problem which the great Sir Isaac Newton is supposed to have contrived after having discussed livestock with an Irish peasant.

109. Newton's feeding problem.

A farmer found that three of his cows could live for two weeks on the grass which stood on two acres of his pasture, plus the grass which grew on this area during the two weeks. He further found out that two of his cows could live for four weeks on the grass that stood on two acres of his pasture plus that which grew on them during this time. How many cows can this farmer feed for six weeks with the grass standing on six acres of his pasture, including whatever grass grows on that area during the time, provided that all of the cows require the same quantity of herbage and that the grass grows at the same rate all over the farm?

110. A variant of Newton's problem.

A farmer owns a cow, a goat and a goose. He finds that his cow needs as much grass as the goat and goose together and that his pasture can either feed his cow and his goat for 45 days, or his cow and his goose for 60 days, or his goat and his goose for 90 days. For how many days of grazing will his pasture last if all three of his domestic animals browse together, taking the growing rate of the grass as constant?

111. Dilemma of the farmer's wife.

A farmer's wife calls on another countrywoman to buy some hatching eggs for her brooding hen. She asks how many varieties of chicken are represented in the other woman's yard. Four different varieties, the other replies, adding that she keeps them rigidly separated but never bothers to keep the eggs distinct. The farmer's wife says she wants to hatch at least five chickens of one single variety, never mind which. The problem is simple, both women think. The farmer's wife just has to put a sufficient number of eggs into her hen's nest, and she will no doubt get five chicks of one variety. And that's exactly as far as they got for they couldn't make up their minds how many eggs would be sufficient. Now, would you know what is the minimum of eggs the farmer's wife must buy to be sure of getting five chicks of the same variety, provided that none of the eggs she buys will spoil?

112. Most favorable utilization of a field.

This simple mathematical problem is not without real practical significance to farmers. What is the most favorable arrangement of plants which have to be planted at a minimum distance from one another in a field? Would a quadratic arrangement make for a maximum of plants on a certain area or is there a better solution? Let's suppose that the field is so large that border strips can be ignored.

113. The missing dollar.

There was once a farmer who drove to the market place to sell his 30 geese at a rate of three for a dollar. On his way he passed the farm of a friend who asked him to take his 30 geese along and sell them too, but at a rate of two for a dollar. The first farmer agreed. However, to make the job easier he decided to make a single flock of both his and his neighbor's birds and to sell every five geese for two dollars. And that's exactly what he did.

On his way home he paid his neighbor the 15 dollars due him. But when he was about to deliver to his wife his share in the sale he realized to his amazement that only nine dollars were left to him instead of ten. Since he was sure he hadn't spent a whole dollar on drinks he cudgeled his brain to find out what had become of the missing dollar. He never did find out, but you might have been able to help the poor fellow.

114. Chopping wood.

A farmer orders his two sons to cut five cords of wood, with the aid of a saw and an axe. The elder son can split logs twice as fast as the younger son, using the saw, can get them ready for him. When the younger son wields the axe, he can split only three quarters of what the other boy can saw in the meantime. How should the two boys distribute their work so they can do the job in a minimum of time?

115. Raising poultry.

A farmer's wife is asked whether her poultry raising business is flourishing. "Oh, yes," she replies, being as good at arithmetic as at raising chickens, "last year I started with 25 chickens and ducks together. Today I have eight times as many; my chickens have multiplied three times as fast as the ducks." With how many chickens and how many ducks did the woman start?

116. Fibonacci's problem.

The mathematician Leonardo Fibonacci, born at Pisa, Italy, in 1175, illustrated an interesting mathematical sequence with the increase pattern of live-stock under certain conditions. This is the problem he devised:

A farmer asked his learned friend: "Suppose that my cow which first calved when in her second year, brings forth a female calf every year, and each she-calf, like her mother, will start calving in her second year and will bring forth a female heifer every year, and so on, how many calves will have sprung from my cow and all her descendants in 25 years?

71

117. Two bellicose goats.

On the center line of a rectangularly shaped meadow measuring 10 by $17\frac{1}{2}$ yards there are two poles with rings, $7\frac{1}{2}$ yards apart, as shown in Fig. 8. Each of the poles is 7 yards from the

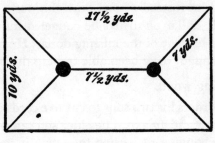

Fig. 8

nearest corners. Two goats, fastened to the poles by ropes through their collars, must be able to graze the entire meadow without ever being in a position to fight each other. In what way should the goats be fastened to the poles and how much rope is needed for the job?

CHAPTER XII

Shopping Puzzles

Buying and selling is always associated with figuring, though ordinarily only with trivial applications of the four basic arithmetical operations—addition, subtraction, multiplication and division. Occasionally, however, the storekeeper or even the buyer may meet with problems which are less elementary; and as we well know from our grade school math texts, the retail business with its measuring, weighing, money changing and price figuring is an inexhaustible source of examples to illustrate ordinary mathematical problems. However, it has always proved a fertile field, too, for the inventive mathematician. You will see, though, that some of the following problems are not necessarily invented. They might well occur in your own experience. Take, for instance, the first puzzle.

118. The counterfeit bill.

A customer in Sullivan's shoe store buys a pair of shoes for $12 and pays with a $20 bill. Mr. Sullivan has no change and walks over to McMahon's neighboring delicatessen to

change the bill. A minute after the customer has left, McMahon storms into Sullivan's store, indignantly waves the $20 bill under Sullivan's nose and tells him it is counterfeit. Of course, Sullivan has to redeem the bill by paying $20. On top of this $20 loss Sullivan had given $8 change to the crook who bought the shoes. So it looks as if Sullivan has lost $28 altogether. At least, that's what he figured. Was he right or wrong?

119. Buying cheese.

A customer comes to an Italian grocer known for his excellent cheese and asks for two cheeses which together will weigh 37 pounds. They must be exactly 37 pounds, the customer insists, because today is his wife's 37th birthday. This, Mr. Marinacci thinks, is rather a peculiar request, but since this is an old and faithful customer he wants to satisfy him. So he take down all his five huge cheeses that he keeps on the shelf and puts first one pair, then another pair on the scales to find two which will satisfy his customer. It is all in vain, however, because every possible combinations of two of the cheeses gave only the following weights: 20, 24, 30, 35, 36, 40, 41, 45 and 51 pounds. So, unfortunately, the customer's wife never got this pungent 37-pound birthday present. You, however, have all you need to figure out how much each of Mr. Marinacci's five cheeses weighed.

120. One pair of socks.

Jimmy Collins, a retailer, had bought an equal number of pairs of black, brown and white socks. The black socks cost him $1 a pair, the brown $1.05 and the white $1.10. Business was bad the next day and Mr. Collins had plenty of leisure to do a little figuring. This is what he found: He could have bought one more pair of socks if he had divided his money equally among socks of the three colors, spending one-third for pairs of each color. How many pairs of socks had Mr. Collins bought the previous day?

121. Christmas shopping.

Four married couples did their Christmas shopping together. Sitting around the dinner table after a strenuous round through all the big downtown stores they figured out how much money they had spent. They found that altogether they were poorer by $500. Mrs. May had spent $10; Mrs. Hull, $20; Mrs. Shelton, $30; and Mrs. Ziff, $40. As to the four husbands, Robert spent five times as much as his wife, Anthony four times as much as his, Henry three times as much as his and Peter twice as much as his. What is the last name of each of the men?

122. Switching the dollars and cents.

James Newman had almost a hundred dollars. So he spent exactly half his money in a department store. When he counted his cash he discovered that he now owned exactly as many cents as he had had dollars before he began spending his money, but only half as many dollars as he at first had cents. With how much money had he entered the store?

123. Buying cigars.

Whenever a cigar smoker replenished his stock, he bought six cellophane-wrapped packages. He had a choice among three different brands. A package of each of the three brands

cost the same, but the cigars were of different quality and so each package contained either three, four or five cigars. He liked to alternate the brands and after a while he found out that with seven purchases he could get seven different quantities of cigars, though all three brands were represented in every purchase. What were the seven assortments the cigar smoker bought?

124. A stamp dealer's dilemma.

A stamp dealer bought five sets of rare foreign stamps, each set showing the face values 15, 10, 8, 5, 4, 3, 2 and 1. Eight of his customers were on the waiting list for this particular set. Having only five sets he couldn't send a complete set to each of them. However, he wanted to distribute what he had as equally and equitably as possible from whatever angle the deal were looked at.

Each of his customers should not only receive the same number of stamps, he reasoned, but moreover, the same amount of face value. Furthermore, the invoices sent to each customer should all be for the same amount. Only the stamp with the face value 2 had a fixed market value of $3. As to the prices of the other stamps of the set, the dealer could fix them more or less freely, charging each customer, of course, the same amount for the same stamp.

In what way should the stamp dealer distribute his five sets and how much would he have to charge—in full dollars —for each stamp so that the invoice amount was the same for each of his customers? (The first part of this problem has several solutions while only one answer to the second question is possible.)

125. Stamp auction.

Two stamp collectors went to an auction because both were interested in a group of three stamps of equal value to be brought under the hammer. Sheridan who couldn't spend more than $50 was soon eliminated from the bidding and so, soon afterward, was Malone who had $70. When the bids had climbed to $110, Sheridan and Malone decided to join forces. So they bid together and acquired the stamps for $120. Each took one of the stamps for his own collection. Some time later they sold the third for $120. How would they have to distribute this sum to make it accord with their shares in the investment?

126. The wooden disk.

To measure ribbons and other material an old-fashioned lady, owner of a New England village store, uses a circular wooden disk fixed on a block on the counter. The disk's circumference is 27 inches. The disk is marked at only six points of its periphery but the marks are placed in such a way that the old lady can measure any length from 1 to 27 inches. Where on the disk are these marks?

127. The unreliable balance.

Some years ago, a grocer acquired a new balance, one of those old-fashioned pairs of scales they used to have in stores before automatic scales were introduced. The first time he used the new balance it turned out that it indicated two different weights, depending on whether he put the merchandise in one scale pan or the other. On this first trial it turned out that the merchandise weighed only $\frac{9}{11}$ of its real weight plus 4 ounces when he put it in one pan and that when it was placed in the other pan it appeared to weigh 48 ounces more than it did in the first weighing. At rest, the balance was at equilibrium, that is, the beam was horizontal. Can you figure the real weight of the merchandise the grocer first put on his faulty scale?

CHAPTER XIII

Whimsical Numbers

Numbers are whimsical. Some catch the eye because of their unusual arrangement or their symmetry, others distinguish themselves by peculiarities that make them freaks, as it were, in the field of the abstract. These peculiarities are often hard to explain. They are sometimes even hard to detect; the oddities are by no means obvious.

Thus, freak numbers of this kind have to be constructed, so to speak, and in most cases the construction is no easy task. It demands skillful use of the basic rules of algebra or the laws of number theory.

To show what we mean, and as a typical example of this type of puzzle, let's try to discover a number, the sevenfold product of which can be found by shifting its last digit to the first place, like this: $7 \times 10,178 = 81,017$ (which, by the way, is wrong). The problem is especially difficult because you are not told how many digits the number in question has.

If you are a trained mathematician you will, of course, proceed methodically: You denote the unit of the unknown number N by y, and the remaining portion of N by x. Then $N =$

$10x + y$ and $7N = 70x + 7y$. If you shift the last digit to the first place the new figure is represented by $y \times 10^z + x$. The value of z is unkown since you do not know how many digits there are in the unknown number. Thus you have the Diophantine equation $y \times 10^z + x = 70x + 7y$ or $x = \dfrac{y(10^z - 7)}{69}$, with three unknowns, x, y and z, which are all integers.

The expression for x with its denominator $69 = 3 \times 23$ indicates that either y is divisible by 3, so that we can write, $y = 3r$, and $(10^z - 7)$ is divisible by 23, or if y is not divisible by 3, $(10^z - 7)$ is divisible by 69, In the first case, $z = 21$, and $(10^z - 7) \div 23 = 99999\ldots\ldots3 \div 23$ begins with 434 and has 20 digits so that $x = r \times 434\ldots\ldots\ldots$ Since x has 21 digits, r must equal at least 3 and can, on the other hand, only be 3 because $y = 3r$ and has to remain below 10. Thus N becomes 1,304,347,826,086,956,521,739.

In the second case, z is again 21, $(10^z - 7) \div 69$ begins with 144.... and, again, has 20 digits. Since $x = y \times 144\ldots\ldots$ has 21 digits, x can be only 8 or 9 so that there are two more solutions for N.

If you are unburdened by so much mathematical equipment, you may reach your goal much more quickly by simple reasoning. You may state right away that under all circumstances the first figure of the unknown number must be a 1, since otherwise a multiplication by 7 would result in more digits than the original number had. The unit of that number can only be 7, 8 or 9, because you need one of the higher numerals, at least 7, at the tail of the unknown number. This is the number which is later to be shifted to the first place, as the result of multiplication by 7 which is not supposed to increase the number of digits.

Try 7 first. In that case, the product will begin with 71..... If you start dividing by 7, you get the first two digits of the original figure, namely, 10..... Therefore, 0 is the third digit of the product. Now you can proceed with the division until no remainder is left. You will find the unknown number to be 1,014,492,753,623,188,405,797. The same result can be obtained by multiplication instead of division. The multiplication method will first yield the sevenfold product of the unknown number.

This example will enable you to solve the following problem.

128. The shifted eight.

A rather long number ends with an 8. If this figure is shifted to the front position we obtain twice the original number.

129. Reversed and fourfold.

Which five-digit number, inverted, equals its fourfold product? There is only one solution to this problem. Oddly enough, the number is twice another figure which, also inverted, equals its ninefold product.

130. Simple division by 3.

There is a multi-digit number which may be divided by 3 simply by shifting the last digit to the front.

131. Division by 4.

Which number, divided by 4, yields itself, except that its unit will be in front? There are three solutions.

132. The wandering unit.

A three-digit number ending with 4 has the following peculiarity: If the 4 is shifted to the first place, the new number is as much greater than 400 as the original number was smaller than 400. What is the number?

133. Sum of combinations.

Which four-digit number is three times the sum of all two-digit combinations that can be made of its digits?

134. Eleven times the sum of its digits.

What three-digit numbers, increased by 11 times the sum of their digits, yield their own inversions?

135. Framed by sevens.

Which three-digit numbers yield products of themselves if supplemented by 7's placed at the beginning and the end?

136. Framed in ones.

There is a seven-digit number which, multiplied by 27, yields itself framed in 1's at the beginning and the end. By framing an eight-digit number by two 1's you will obtain the number multiplied by 29. Finally, there is a ten-digit figure which, multiplied by 33, yields itself framed in two 1's. What are these three numbers?

137. Prime or no prime?

Is the number 1,000,000,000,001 a prime or not? If it is not a prime, what are its two components?

The questions would be no more difficult to answer if the number had 101 zeroes instead of the 11 zeroes given above. All numbers built that way have certain peculiarities.

138. A reversible multiplication.

Look at the product $2,618 \times 11 = 28,798$. If you invert the first figure and again multiply by 11 you will obtain a product that is the inversion of the first product: $8,162 \times 11 = 89,782$. Can you find another number with the same peculiarity?

139. All numerals from one to nine.

Both of the two products $12 \times 483 = 5,796$ and $42 \times 138 = 5,796$ contain all numerals from 1 to 9 and moreover, yield the same result. Can you find another such pair of products with the same peculiarities?

140. All numerals from 1 to 9 yield 100.

In the following addition of two positive and one negative numbers, the letters A to I shall be replaced by the numerals 1 to 9 so that the sum is 100:

$$
\begin{array}{r}
+\ A\ B\ C \\
+\ D\ E\ F \\
-\ G\ H\ I \\
\hline
=\ 1\ 0\ 0
\end{array}
$$

141. Raising two numerals.

In a four-digit number the numerals in the hundred and units position are to be raised so that they indicate the powers

81

of the same numerals in the thousand and ten position, and the product of the two resulting powers are to equal the original number. The numerals 1 and 0 are not supposed to be used as the figures indicating the powers. For instance, $3^4 \times 7^2 = 3,472$, which, by the way, is incorrect. Can you find the four figures in question?

142. Reducing a fraction.

If we delete the numeral occurring in both the numerator and the denominator of the fraction $\frac{26}{65}$ we don't change its value because $\frac{26}{65} = \frac{2}{5}$. What other fractions consisting of two-digit figures in the numerator and denominator can be similarly reduced?

143. Several ways of expressing 1000.

In what way can 1000 be expressed by the sum of two or more consecutive numbers?

144. Phoenix numbers.

The difference between a number and its inversion may result in a new number which contains the same numerals as the old, though in a different order. Take, for instance, the following subtraction in which the numerals change their places in accordance with the alphabetical pattern at right:

$$
\begin{array}{r} 954 \\ -459 \\ \hline 495 \end{array}
\qquad \text{or} \qquad
\begin{array}{r} ABC \\ -CBA \\ \hline CAB. \end{array}
$$

This kind of play with numbers can be extended to larger numbers and you may try to construct similar subtractions of five- and seven-digit numbers which obey the same rule, as

$$
\begin{array}{r} ABCDE \\ -EDCBA \\ \hline ABCDE \end{array} \text{ in a different order}
\qquad \text{and} \qquad
\begin{array}{r} ABCDEFG \\ -GFEDCBA \\ \hline ABCDEFG \end{array} \text{ in a different order.}
$$

Finally, you may try to find an analogous nine-digit subtraction containing all numerals from 1 to 9,

$$
\begin{array}{r} ABCDEFGHI \\ -IHGFEDCBA \\ \hline ABCDEFGHI, \end{array} \text{ differently arranged.}
$$

This latter problem has an amazingly large number of solutions.

CHAPTER XIV

PLAYING WITH SQUARES

This chapter deals with squares, that is, squares in the arithmetical sense. You can have a lot of fun with squares if you choose to delve a little into this branch of mathematics. On the other hand, you may need some theoretical knowledge of this particular branch of mathematics for solving some of the square problems. And, of course, you won't get very far without an acute sense for figures. You will find that these puzzles cannot be solved by a single fixed line of march. They are all more or less different, and in practically every case you will need an inspiration or two to reach the goal.

Moreover, you should keep in mind the specific peculiarities of square numbers. For example, any odd number can be expressed in one or several ways by the difference of two squares; if the number is a prime, there is only one such way (for instance, $17 = 9^2 - 8^2$); if not, there are at least two (for instance, $15 = 8^2 - 7^2 = 4^2 - 1^2$). Any number divisible by 4 can be expressed by the difference of two squares (for instance, $4^2 - 2^2 = 12$.) If a square is divisible by a prime, the square of the prime, too, divides into the square. If the unit of a four-digit square is 5, it invariably ends in 025, 225 or 625. If a square ends on 6, its next to the last digit is odd; in every other case it is even. The sum of the digits of a square or else the sum of the digits of the resulting number, or the sum of the digits of that number, etc. can only be 1, 4, 7 or 9. (For example 49: $4 + 9 = 13$, $1 + 3 = 4$.)

In many of the following problems you will find amazingly few statements on the numbers involved. But the simplicity of the question in no way implies that its solution is so simple. Take for instance, the following problem which is particularly difficult to solve and for which, therefore, we shall supply you with a number of hints.

145. Crossword puzzle without words.

Into the boxes of this "crossword puzzle" (Fig. 9) one-digit numerals are to be inserted in such a way that all the resulting

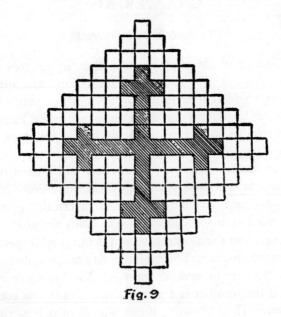

Fig. 9

horizontal and vertical numbers are squares. To make things somewhat easier, we shall give you some correct numbers and a number of clues. Into the uppermost box goes a 4. To fulfill the conditions of the problem, either 1, 4 or 9 must go into the three other corners. The second horizontal number is 441 (= 21^2), the third horizontal number is 68121 (= 261^2). Consequently, the central vertical column at the top gets again the square number 441. In the vertical column to the right you have 1 and 2 as the first two numerals, which means that the third number must be 1 and the figure is 121 (= 11^2). For the vertical column to the left of the center top column you will quickly find the number 484 (= 22^2), and so on.

146. The inverted square.

What four-digit square, when it is inverted, becomes the square of another number?

84

147. Squares composed of even numerals.

Find all four-digit squares composed exclusively of even numerals. There are four solutions altogether.

148. Square of the sum of its parts.

Find a four-digit number that equals the square of the sum of the two two-digit numbers formed by its first and second, and third and fourth numerals, respectively.

149. Inversions of roots and squares.

The square of 12 is 144. The square of 21, the inversion of 12, is 441, which is the inversion of 144. This means that the equation $12^2 = 144$ allows the inversion of the numbers on both sides of the sign of equality. The same thing is true of 13 and its square; $13^2 = 169$, $31^2 = 961$. 12 and 21, and 13 and 31, are the only pairs of two-digit numbers that satisfy this requirement.

We now ask, can three-digit numbers also satisfy this requirement? They can, and we may add that there are five such pairs of three-digit numbers. What are they?

150. A shortcut.

Do you know a quick way to decide whether the number 15,763,530,163,289 is a square?

151. A square consisting of two consecutive numbers.

There is one four-digit square number, the first two digits of which equal the last two digits plus 1. For instance, 5857 might be this particular square number, but isn't. What number is it?

152. An odd set of squares.

If we insert into the square number 49 the number 48, the new number 4489 is another square, namely 67^2. The same holds true for all numbers formed from 4489 by inserting more 4's and 8's, such as 444889, 44448889, and so forth. What is the reason for this peculiarity? Furthermore, are there more two-digit numbers of this kind?

153. A combination of squares.

What two-digit squares may be combined into a four-digit square?

154. Squares around a triangle.

Fig. 10 shows an equilateral triangle with the numbers 1 to 9 arranged along its sides and corners. Try to find an arrange-

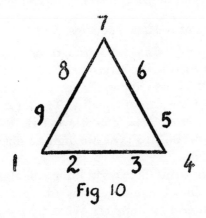

Fig 10

ment such that the sums of the squares of the four numbers along each side of the triangle are equal.

155. Odd series of consecutive numbers.

Find a series of consecutive numbers, the sum of which will equal the product of the first and last number in the series.

To solve this problem you will have to recall the laws governing Pythagorean number triples. You may remember that any triple of integers answering the equation $x^2 = y^2 + z^2$ is called a Pythagorean number triple. The lowest number triple answering this equation is 3, 4 and 5. The formulas $y = a^2 - b^2$, $z = 2ab$ and $x = a^2 + b^2$ yield all primitive Pythagorean number triples x, y and z, that is, those triples which have no common factor. a and b must be positive integers, a must be greater than b, a and b must have no common factor and must not both be odd. With the aid of the above three equations each pair of integers satisfying these conditions yields three numbers which are primitive Pythagorean number triples.

Equipped with this knowledge you will be able to solve the problem which yields a special class of Pythagorean number triples.

CHAPTER XV

MISCELLANEOUS PROBLEMS

In this chapter you will find a number of puzzles which don't really fit into any of the other chapters of this book. Some are rather new, some are very ancient and for a long time have been part and parcel of mathematical entertainment. The following is one of these staple problems.

156. Pythagoras' thievish slave.

Pythagoras once punished a thievish slave, whom he had caught redhanded, by ordering him to walk up and down past the seven columns of the temple of Diana, counting them, till he reached the one-thousandth column. The columns were in one line and the slave was supposed to count them by walking from the left to the right, turning around at the seventh column, counting the sixth as the eighth, then again turning around at the first, now counting it as the thirteenth, the second one as the fourteenth and so forth. After having finished his march, he was supposed to report to his master and tell him which column he had counted as number 1,000. Of course, the great mathematician Pythagoras had figured out which of the seven columns would be number 1,000 and thus was in a position to know right away whether the slave had followed his rather foolish order. If you are as smart as Pythagoras you won't have much trouble finding out at which column the poor slave's ordeal was over.

157. Ancestor number 1,000.

This is another puzzle involving the number 1,000. There was a man in New Orleans who traced his ancestry back to an impressive number of personages of royal blood and other aristocrats. His family tree, drawn on canvas, covered an entire wall and was very confusing, especially when this descendant of all these dukes and princes searched for a particular member of his family dead for hundreds of years. So he devised a clever system for listing and finding any of his ancestors. He gave himself the number 1; his father the number 2; his mother, 3; his four grandparents, 4 to 7; his eight great-grandparents, 8 to 15; and so on.

Using this system of tabulation, the mother of the man's maternal grandfather, for instance, got number 13. Now, which number had the mother of the mother of the father of the father of the mother of the mother of his father?

We ask this question only to give you a chance to get familiar with this peculiar system of pedigree registration. As to the question involving the number 1,000, the man with the illustrious ancestors found that this figure was already reached by the ninth generation back. Can you find out what the relationship was between our contemporary and the ancestor number 1,000?

158. A question referring to an exam.

A number of boys and girls had flunked a comprehensive language examination because they did badly in one or more of three subjects, English, French and Spanish. The teachers were discussing the poor results of the semester's hard labor and one of them, looking over the list, remarked: "Two-thirds of those who flunked failed in English, three-fourths in French and four-fifths in Spanish." "From that I see," interrupted the mathematics teacher, glancing at the list, "that 26 of those who flunked failed in all three items and, moreover, that this is the smallest figure possible. The other teachers found out that the mathematician was right. Can you, on the basis of these figures, compute the number of boys and girls that flunked?

159. A one hundred per cent sure tip.

Fred Robinson had a one hundred per cent sure tip on the horses. Somebody had told him that one of four outsiders was bound to win. On the first of these, the odds were 4 to 1, on the second 5 to 1, on the third 6 to 1 and on the fourth, 7 to 1. Fred is a modest fellow and didn't want to make more than $100 on this tip. However, there being a 1% state tax on net winnings, Fred would have to make $101 more than he staked on the race. How much would Fred have had to wager on each of the four horses to have a profit of $101, no matter which of the four outsiders should win?

160. How many children in this family?

A man with quite a numerous family was once asked how many children he had. "Plenty," he replied, "and moreover, I have found out that even my smallest children eat just as much of everything as my wife and I do. After we were blessed with our latest family increase, the twins, I found out that our crate of oranges lasted three days less than before that event. If I had another four children, that crate would be empty even four days earlier than it is now." How many children did that lucky man have?

161. Three hoboes, three loaves and one dollar.

Once three hoboes met on the highway. One had three loaves of bread, the second two and the third none, though he owned a fortune of one dollar. They decided to distribute their edible assets in equal parts. The third hobo, of course, had to compensate his colleagues out of his dollar. How much did he give to each of the bread owners?

162. To each his own.

This is one of the old stand-bys of mathematical entertainment. John of Palermo in the presence of the German Emperor Frederick II is supposed to have set Leonardo of Pisa the task of solving the following puzzle:

Three men collectively owned a certain number of ducats. Their shares were one half, one third and one sixth, respectively. They made a pile of all the coins and each of them took a part of it so that none of the money was left. Then the first returned one half; the second, one third; and the third, one sixth of what he had grabbed and finally, each of the three got an equal share of what had been returned by all three put together. And then each had exactly the number of ducats that really belonged to him.

What is the smallest number of ducats with which this transaction will work and how much did each of the three men grab from the pile in the first place?

163. The two towers.

This problem is supposed to have been devised by Leonardo of Pisa mentioned in the preceding problem. Two towers, one 30, the other 40 rods high, are 50 rods apart. There is a milestone on the straight line between these towers. From the tops of both towers two crows fly off simultaneously and with the same speed in a straight line in the direction of the milestone and—believe it or not—reach their goal simultaneously. How far is the milestone from both towers?

164. The hymn book.

An old hymn book contains 700 hymns, numbered 1 to 700. Each Sunday, the church choir sings four different hymns. The numbers of these hymns are made known to the congregation by combining single number plates on a blackboard. Now, what is the minimum number of number plates required for composing any possible number combination of the four hymns, provided that the plate for 6 may be turned upside down to serve as a 9?

165. The mathematical-minded philanthropist.

Sometime in the thirties, before 1939, a philanthropist decided to distribute his fortune by giving, each day, throughout the next year, the same amount to the poor. He had found out that his fortune could be disposed of without remainder, if every day of that year he should distribute a certain full number of dollars (no cents). But if he should distribute his fortune in even dollars at a different rate and only on weekdays, one dollar of it would remain at the end of the year. The generous old man owned exactly the minimum amount which empowered him to use it in the way he wanted. You, on the other hand, by what you have learnt of the philanthropist's good intentions, are in a position to determine not only the amount of dollars the good man owned but also the year in which he came to his resolution.

166. McMillan's savings.

McMillan's weekly paycheck is one half greater than that of his colleague, Barber. Barber spends only half of what McMillan is used to squandering. If McMillan saves only half as much of his wages as Barber does, what is the ratio of McMillan's disbursements to his salary?

167. A handsome ransom.

After a sea battle a pirate captain fell into the hands of a Persian prince. This generous prince, who was well versed in the art of figuring, did not want to execute the pirate without giving him a chance to save his life. So he asked him to deliver, within ten days, one million silver coins, in accordance with the following schedule: On the first day, he had to deliver more than one hundred pieces of silver. On the second day, he again had to deliver more than one hundred coins. On the third day, he would have to pay as much as on the second day and on top of that as much as he had paid on the two previous days. On the fourth day, he would have to pay as much as on the third day plus the rates delivered on each of the previous days including the third day, and so on, up to the tenth day. The sum total of all silver coins paid up to that date would have to be exactly one million.

So the pirate captain was left alone in his dungeon to find just the right sums to be delivered on the first and second days so that, following the prescribed schedule, he would reach the million mark just in time. Well, fortunately, he, too, was well versed in the art of arithmetic and moreover, he owned a considerable treasure in silver coins so that he could buy his freedom. How big were the two initial rates?

168. Up the Hudson River by steamer.

On a beautiful spring day, a steamship went up the Hudson River from New York to Albany. When the ship left New York it was pretty well filled. At Bear Mountain Park, half of the passengers left the ship and 12 new ones came aboard. At West Point, again half of the passengers went ashore and ten new ones boarded the vessel. At Newburgh, once more half of the people aboard left while eight passengers were taken on. At Poughkeepsie, again half the passengers went ashore and six new ones came on board. At Kingston, again fifty per cent of the passengers left while only four new ones came aboard. At Catskill, again half of those aboard left but just two new ones came aboard. Finally at Albany the last ten people went ashore. How many passengers were aboard when the ship left New York?

169. Vote with obstacles.

A vote had to be taken at a crowded club meeting and the president suggested that all who were in favor of the motion should stand while those in opposition should sit down. After a count had been taken, the president stated that the proposal had been accepted with a majority count greater by exactly one quarter of the opposition vote. Thereupon there were protests by club members insisting that the vote was void because there had not been enough seats for all the opponents of the motion. Twelve members stated that they had to stand against their wills. Thereupon the president made known his decision that under these circumstances the motion had been rejected by a majority of one vote. How many people were present at the club meeting?

170. Marriage statistics.

Of the inhabitants of a small European republic, 4.2 per cent of the male inhabitants and 2.8 per cent of the female population married mates of their own nationality within one year. What is the ratio of the male to the female inhabitants of this state?

171. Where is the mistake?

A division problem given in school resulted in a quotient of 57 and a remainder of 52. One student who had solved the problem wasn't quite sure of his figuring efficiency and so he made a test by multiplying the quotient with the divisor and adding the remainder. The result was 17,380, which unfortunately was in no way identical with the original dividend. The reason was that the student when doing the multiplication had mistaken a six, the second figure of the divisor from the right, for a zero. What was the original division problem?

172. Mail trouble.

In a certain war-torn country, the postal service is far from having returned to normal. On the average, one sixth of all letters do not reach their destinations. In that country, a boy studying at a university in the capital wrote a letter to his father, who had always answered every letter promptly. He got no answer. What is the probability that his father never got the letter? What is the probability that, excluding unexpected events such as the father's death and assuming the received letter has been answered, the son would receive the answer? That is, generally speaking, what is the ratio of all answers received to the number of all letters sent?

173. Just distribution of marbles.

Seven boys owned a certain total number of marbles but there was considerable difference of opinion as to how many belonged to each. The father of one of the boys, a mathematics teacher, decided to settle the matter once for all and to everybody's satisfaction. So he asked each of the children to get a basket. Then he distributed the marbles in a peculiar way which looked far from just. Finally, he asked the first child to dip into the marbles in his basket and put into each of the others as many marbles as were already in it. Then he asked the second boy to do exactly the same thing, and so on, up to the seventh boy. When he asked each of them to count the number of marbles now in his basket, they found that each had 128. How many marbles had the father put into each of the seven baskets?

174. Cigarette paper.

Paper to be fed to cigarette machines comes in long bands wound in a tight roll around a wooden spool. The diameter of such a paper roll is 16 inches, that of the spool itself, 4 inches. Would you know how long the paper band wound around the spool is if the paper is $\frac{1}{250}$ of an inch thick, provided there is no measurable space between the paper layers?

175. Involved relationship.

There are innumerable problems concerning relationship of two or more people, most of which, however, have little to do with figuring or mathematical entertainment. We will therefore present only one such puzzle, and this only because of its historical interest. A collection of puzzles called *Annales Stadenses*, published in Germany in the XIIIth century, contains the following problem:

Each of two women had a little son in her arms. When someone asked them how they happened to have these sons they answered: "They are our sons, sons of our sons and brothers of our husbands and everything is strictly legitimate." How was that possible?

CHAPTER XVI

PROBLEMS OF ARRANGEMENT

In these problems of arrangement we shall not deal with games such as chess, checkers, halma or others in which the winner defeats his adversary by a more ingenious arrangement of his pieces or men. In this chapter we are interested only in problems which permit a more or less mathematical treatment.

To be sure, many such problems are rather close to the domain of games. Some of these, such as the Problem of the Knight, the Problem of the Five Queens, the Fifteen Puzzle, and so on, have been considered, because of their intermediary position, worthy of scientific treatment by great mathematicians. Thus, in 1758, Leonhard Euler submitted to the Berlin Academy of Sciences a paper on the Problem of the Knight. The most prominent problem of arrangement since ancient times has been the magic square. It fascinated great thinkers in the India of Sanskrit times and is by no means scorned by modern mathematicians. Many of you are probably already familiar with the structure of a magic square. It is a square array of n times n distinct positive integers, often consecutive ones, arranged in such a way that the sum of the n numbers in any horizontal, vertical or main diagonal line is always the same. Construction of such magic squares has developed into a real sciences and ingenious methods have been devised for their formation.

However, we doubt that the whole magic squares business is worth all the time and labor spent on this pastime. We have, therefore, no intention of describing the methods of their construction, the less so because once any of the well-known methods for solving them are understood, only a mechanical application of these methods is required. However, there are a number of quite different problems of arrangement which are rather interesting when looked at from a mathematical point of view.

Let us begin with a comparatively simple problem, which is just on the borderline between mere entertainment and science. It was first presented to the public and theoretically investigated by the well-known physicist Taits around the turn of the century.

There are four white and four black pieces alternately arranged in a row, as indicated in the first line below. Two empty spaces are left at the right. The problem is to shift these eight pieces in only four moves so that the white and black pieces, respectively, are lying together, as shown in the last line below. Moves are made only by shifting two adjacent pieces simultaneously, without changing their position relative to each other. This is the solution of Taits' problem:

```
W  B  W  B  W  B  W  B  *  *
W  *  *  B  W  B  W  B  B  W
W  W  B  B  *  *  W  B  B  W
W  W  B  B  B  B  W  *  *  W
*  *  B  B  B  B  W  W  W  W
```

These moves are easy to keep in mind, if the fourth piece, which is never moved, is taken as a fixed point. The pair to the left of the fixed point is moved over to the right and the pair to the right of the fixed point fills the resulting gap. The two remaining moves are made by pairs of the same color.

This problem may be expanded, leading to the following variant:

176. Ten pieces.

There are five white and five black pieces alternately arranged in a line. Otherwise, the problem is the same as the one just explained. Five moves should do the job.

177. Eight coins.

Eight coins are lying in a straight line. By jumping one coin over two others, whether stacked or adjacent, in each move, so that it rests on top of a third, how can you finally get all the coins stacked in pairs?

If you are especially ambitious you may try a longer line of coins. The number of coins, however, should always be even.

178. The problem of the 16 cards.

The great mathematician Leonhard Euler hardly neglected any branch of mathematical entertainment provided he was sure to encounter plenty of difficulties. There are only a few such problems which Euler was unable to solve. One of them was the arrangement of the 36 officers.

These 36 officers belong to six different regiments and have six different ranks, and each regiment is represented by one officer of each rank. The 36 officers are supposed to be arranged in lines of six each, so that officers of neither the same regiment nor the same rank stand in the same horizontal or vertical line.

Euler was convinced that the problem was insoluble, though he also thought rigid proof of its insolubility to be very difficult. Not until the nineties of the last century did a mathematician, Tarry, succeed in proving that a solution is impossible.

Strangely enough, Euler squares—as such problems are called—of a higher order than 6 (and also of a lower order) are often soluble. For example, 64 officers of 8 different regiments and 8 different ranks have no difficulty arranging themselves as requested by Euler.

We will refrain from giving the entire theory of the Euler squares and present only one such problem, a simple one at that, which may be solved by nothing other than a little instinct for combination.

Ace, king, queen and jack of all four suits of a deck of cards are to be arranged in a square in such a way that neither a suit nor a value occurs twice in any of the four-card horizontal or vertical rows.

179. The pentagram.

The points of intersection of the pentagram, Fig. 11, are marked by small circles and the letters A to J. The problem is to write ten different numbers into the circles so that the sum of the four numbers along each of the five lines of the pentagram is 24.

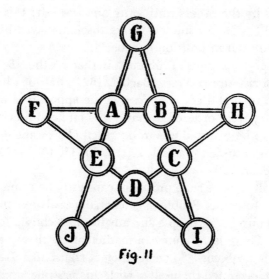

Fig. 11

It can be proved that it is impossible to solve the problem with a series of consecutive numbers. After you have discovered the law governing this problem you may choose any even number as the sum of the numbers along each line of the pentagram, and insert the ten numbers according to the rules you have found, although in many cases you cannot do without negative numbers and duplications of numbers.

180. The multicolored dice.

Imagine a die with each of its six faces painted in any of six different colors. In how many different ways may the six colors be arranged?

181. One of many "Josephus' Problems."

We want to present only one of the type of problems of which Josephus' is the best-known. According to Hegesippus' *De Bello Judaico*, the famous Jewish historian, Josephus, and 40 other Jews were hidden in a cave surrounded by Roman warriors. None of Josephus' companions was willing to fall into the hands of the Roman conquerors, and they resolved to kill each other instead, if necessary. Josephus didn't like the idea but feigned agreement and proposed that all 41 men arrange themselves in a circle. Then every third man was to be killed by the others until only one was left; this one was supposed to commit suicide. Thus Josephus saved his life by placing himself in position number 31.

Another "Josephus' Problem" is that of the 15 Turkish and 15 Christian prisoners (Bachet, 1624), half of whom were supposed to be killed. To preserve the appearance of impartiality they were ordered to stand in a circle and every ninth prisoner to be counted was to be killed. To escape death, the Christians occupied places 1, 2, 3, 4, 10, 11, 13, 14, 15, 17, 20, 21, 25, 28, and 29.

The solutions to such problems can easily be found empirically, and Josephus' Problems can equally easily be constructed with any number of people and any number chosen for counting out. There is, of course, a mathematical theory that governs such problems. Some mathematicians, like Bache and Schubert, occupied themselves with this pastime, and an elaborate literature deals with Josephus' Problems.

However, we do not intend to present any part of the involved theory and want to give you only one such problem, which is simple to solve empirically.

Get one each of all U. S. coins: a penny, a nickel, a dime, a quarter, a half dollar and a silver dollar and arrange them in a circle, in that order. Then tell a friend he may have all six coins, that is $1.91, if he is able to choose the right number for counting out the six coins, beginning with the penny, so that the silver dollar is the last to be counted out. For instance, if you chose the number 7 you would first count out the penny, second the dime, and third the dollar. Thus, 7 is no good. What number, or numbers, can earn your friend $1.91 in two minutes?

182. The ten-spoke wheel.

Imagine a wheel with ten spokes, with a small tablet at the end of each spoke (Fig. 12). Four of the tablets, A, B, F and G, carry the numbers 22, 19, 26 and 13, in that order.

These four numbers are peculiar in that the sum of the squares of two adjacent numbers equals the sum of the squares of the two opposite numbers ($22^2 + 19^2 = 13^2 + 26^2$). You

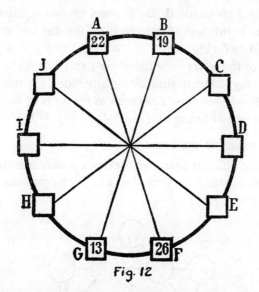

Fig. 12

are supposed to fill the remaining tablets with other numbers in such a way that the same scheme is maintained all around the wheel.

183. The six-pointed star.

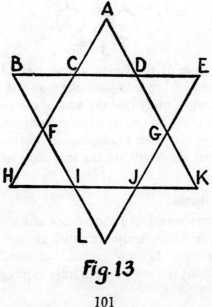

Fig. 13

You have a six-pointed star formed by two equilateral triangles. The points and intersections of the star are marked by the letters A to L (Fig. 13).

Instead of the letters, the numbers 1 to 12 are to be inserted in such a way that the sum along any side of a triangle, for instance, $B + C + D + E$, as well as the sum of six numbers forming the inner hexagon, $C + D + G + J + I + F$, is 26.

184. The seven-pointed star.

Still more difficult than the previous problem is that of the seven-pointed star shown in Fig. 14. (Disregard dotted lines for the moment.)

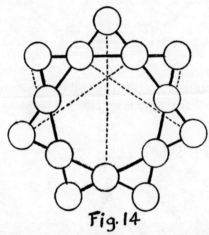

Fig. 14

The seven lines of the figure intersect at 14 points altogether, and the problem is to place the numbers 1 to 14 in the 14 circles so that in every case the sum of the numbers along a line adds up to 30. So far, no strictly mathematical solution of this problem has been found, though certain rules have been discovered that facilitate the empirical solution of the problem.

185. Tchuka Ruma.

Tchuka Ruma is an East Indian game which can be played with the most primitive equipment, such as a number of holes and some pebbles. It may be mathematically analyzed, although its theory has not yet been fully explored.

You have a board with five holes in a row. There are two pebbles (or pieces) in each of the first four holes, while the last, called Ruma, is empty:

I	II	III	IV	R
2	2	2	2	0

The object of the game is to put all the pieces in the Ruma hole in the manner prescribed. The game is begun by taking two pieces from any one hole and placing them, singly, into the next two holes to the right. If you still hold a piece—or, in later moves, more than one—in your hand after you have dropped one in the last hole, the Ruma, you put the next piece in the first hole, at the far left, and from there proceed in your distribution to the right, as usual. If the last piece is dropped in the Ruma you may select any hole for the next distribution. Otherwise you are supposed to empty the hole into which you dropped the last piece, provided it already contains one or more pieces. You may not empty the Ruma. If the last piece out of a hole happens to go into an empty hole, the game is lost. The game is won, on the other hand, when all 8 pieces are in the Ruma.

CHAPTER XVII

PROBLEMS AND GAMES

Of the many problems which can be built around games of all sorts we shall exclude all those which have to do with the theory of probabilities. Questions like that of the probability of a certain card distribution being duplicated have been asked too often to be worth repeating. In many such cases, the amateur mathematician will be stupefied by the tremendous size of the resulting figure. But, as the mathematically trained know, the solution requires nothing but the application of a formula fitting the case in question, and hardly ever involves even a dash of ingenuity.

Aside from questions concerning probability, card games, games of chance—or any other games, for that matter—are fertile soil for arithmetical problems. The following problem may even have some practical significance for card players.

186. Accounting at the card table.

Three persons, A, B and C, play a card game in which the winner is to receive a certain amount from each of the two losers. Ordinarily, the accounting is done according to Schedule I, that is, in such a way that the last line indicates the up-to-the-moment score, whether gain or loss, of each player. However, the accounting can also be done according to Schedule II. Only the winnings are noted, and at the end of the

	Schedule I			Schedule II		
	A	B	C	A	B	C
A wins 40	$+80$	-40	-40	40		
B wins 60	$+20$	$+80$	-100		60	
C wins 30	-10	$+50$	-40			30
A wins 45	$+80$	$+5$	-85	45		
C wins 15	$+65$	-10	-55			15
				85	60	45
				$=sA$	$=sB$	$=sC$

evening, the score of each of the players has to be computed. The question is how to compute the score from the sums sA, sB and sC in Schedule II.

Imagine another card game with four players, A, B, C and D. Two of them play against the other two and the partners change each round. The winnings are equally distributed between the two winners, and the same goes for the losses.

Here, too, the accounting can be done according to Schedule I or the much simpler Schedule II. How would you compute the score at the end of the game from the sums sA, sB, sC and sD?

	Schedule I				Schedule II			
	A	B	C	D	A	B	C	D
A and C each win 120	+120	−120	+120	−120	120		120	
A and D each win 80	+200	−200	+ 40	− 40	80			80
B and C each win 45	+155	−155	+ 85	− 85		45	45	
A and B each win 85	+240	− 70	0	−170	85	85		
B and C each win 90	+150	+ 20	+ 90	−260		90	90	
A and C each win 35	+185	− 15	+125	−295	35		35	
B and D each win 100	+ 85	+ 85	+ 25	−195		100		100
					320	320	290	180
					=sA	=sB	=sC	=sD

187. To cover the table with cards.

Somebody tries completely to cover his rectangular card table, the width of which exceeds its length by 2 inches, with a deck of 52 2 × 2½-inch cards, so that the cards just touch each other. He almost succeeds, leaving an uncovered space of only 4 square inches. However, several cards are left over, consisting of an equal number of clubs and hearts. How many clubs were on the table?

188. A remarkable game of chance.

Six gamblers play a remarkable game of chance. The game itself is rather primitive but the loser is in a bad spot. He is supposed to double the pool of each of the other five gamblers.

Altogether, they play six games and by chance each of the men loses just once.

When the men later count how much is left to each of them they discover that each owns exactly $64. How much had each of them when they started? As with most such puzzles, the easiest way to get the result is to figure back from the end.

189. How many games?

This puzzle is so simple that it almost fits better among the trifles in the first chapter. Two men play a card game, the stakes being $1. When they have finished, one has won three games, while the other has a net profit of $7. How many games did they play?

190. The roulette player's dream.

An ardent roulette player once dreamed that he would soon make $2,100 at his favorite game. He distinctly saw himself in his dream simultaneously staking four piles of chips on certain parts of the board. He put one pile on a single number (plain), another on a double numeral (cheval), and two on each of two groups of six numbers (transversal). The odds for plain, after deduction of the stake, are 36 to 1; for cheval, 18 to 1; and for transversal, 6 to 1.

When the gambler awoke he remembered how he had disposed his stakes, but could not recall their amounts. So he decided to figure what they would have been to make it possible for him to win $2,100, no matter which of the four choices had brought him his triumph, and only one of them could cover the winning number. That is, the single number he chose could not be in any of the groups of six numbers or the other combinations. What were his stakes?

191. Mr. Smith and his children.

In a certain simple game of chance, the first 10 cards (ace, two...ten) of a suit are distributed among five players so that each gets two. Winner holds the highest hand. Once Mr. Smith played this game with his four children. Mary had 11 points, Betty 7, Henry 5 and Charlie 14. What were the hands dealt Mr. Smith and his four children?

192. The remainder for the innkeeper.

Four gamblers played a game of chance in which the pool was less than $1,000. They agreed that at the end of each game the amount in the pool was to be divided by four. If, after this division, a remainder was left, it was to be given to the innkeeper for letting them gamble in his place. The winner was to get one-fourth of the pool and three-fourths was to remain for the next game. The winners in the second, third and fourth games were to be rewarded in the same way, and the innkeeper was to get the remainder, if any. After the fourth game, each of the gamblers was also to get one-fourth of the remaining pool (i.e., the pool which would have started a fifth game).

The gamblers won one game apiece and the innkeeper got $1 after each game. In no case did a fraction of a dollar occur in any of the divisions. What did each of the four players have at the end of the evening, and what was the pool when they started?

193. A sure way to win.

Ask a friend of yours to play a simple game of chance with you. Take 16 cards—the four aces, twos, threes and fours—and arrange them in a rectangle so that at the top you have the aces (each counting one), then the twos, etc. Alternately, each of you will turn one of the cards face down and will note its value. The one who first reaches 22 or compels his partner to go beyond that mark has won. There is a method by which you can always win, provided you turn the first card. How?

194. No profit in gambling.

While Earl was in Reno to get a divorce he patronized a gambling house, though with little luck. Whenever he went there he had to pay a fee of $1, and he gave a tip of $1 to the hat-check girl on leaving for the evening. Well, the first day, Earl lost half of the money left after the entrance fee was paid. The same thing happened the second, third and fourth days. Then Earl had to give up for the very simple reason that but one single dollar was left of his fortune. How large had his fortune been?

195. The bridge tournament.

Four married couples played a bridge tournament which lasted three evenings and was played at two tables. The grouping was changed every day, in such a way that none of the men had the same woman more than once as partner or more than once as opponent. Moreover, no husband and wife ever played at the same table. In what way were the four married couples distributed?

196. Win or lose $200.

Three wealthy couples, the husbands called John, Eric and Otto, the wives (not necessarily in the same order), Ann Beatrice and Margie, visit a gambling ship off the California coast. Each of the six gambles on his or her own account. However, they agree that each couple will stop whenever their joint gain or loss has reached $200. Unfortunately, all three husbands lose all the time, but thanks to their wives' luck, eventually each of the three couples wins the agreed maximum of $200. So everybody is quite happy and when they discuss the events of the evening it turns out that by chance each of the six people has participated in as many single games as, on the average, he or she has won or lost dollars per game. Moreover, they figure out that Eric has lost $504 more than John, and that Beatrice has won $2,376 more than Margie.

These facts are sufficient for you to find out who is married to whom, in case you are interested.

SOLUTIONS

SOLUTIONS

1 – The empty bottle costs $2.50, the wine $42.50, that is, $40 more than the bottle. If you prefer an equation: $b + (b + 40) = 45$; $b = 2.50$.

2 – The dealer was right. The car that brought him a 25 per cent profit cost him $600 (because $600 + 25\% = 750$). The other car on which he lost 25 per cent cost him $1,000 (because $1,000 - 25\% = 750$). So the dealer paid a total of $1,600 for two cars which he sold for $1,500, causing him a loss of $100.

3 – Exactly $111\frac{1}{9}$ pounds. ($111.111\ldots - 10\% = 100$)

4 – Charley thinks he will be in plenty of time when his watch shows 8:20, while in fact it is already 8:30, much too late to make the train. Sam, on the other hand, will reach the station when his watch shows 7:45, which means that he will be there at 7:40, or fifteen minutes too early.

5 – Four brothers and three sisters.

6 – One may be inclined to figure that since the snail manages to gain 5 yards (7 – 2, that is) in every 24 hours, it will reach the edge of the well at the end of the fourth day. This, however, is a fallacy and, moreover, an underestimate of the snail's speed. For it is evident that at the end of the third day, the snail will be 15 yards up. On the fourth day, with a speed of 7 yards, it will reach the rim of the well within five-sevenths of the daylight hours of that day.

7 – It will drop for the following reason: While still in the boat, the cobblestones displace an amount of water equivalent to their weight. Immersed, however, they will displace an amount of water equivalent to their volume, which is less, since cobblestones have a higher specific gravity than water. (Otherwise they wouldn't sink to the bottom.)

8 – It will turn twice because its motion is a composite one, consisting of its own rotation around its center and its rotation around the stationary gear wheel. Try it out with two coins of the same size.

9 – You will get the wrong result if you write this number 11,111. It should be written 12,111, and this number is divisible by 3, because the sum of its digits is divisible by 3.

10 — Only 5, of course (unless they get tired before the job is done).

11 — We really hope you didn't fall into this trap. The number of grooves per inch has nothing whatsoever to do with the problem. The needle doesn't travel around the record; it is the record that turns. The needle is stationary except for its movement toward the center of the disc. That means, it travels $6 - (2 + 1) = 3$ inches. Half the diameter less half inner blank plus outer blank.

12 — The first man would in any case claim that he is a Betan, even if he is an Atan, because in the latter case he is bound to lie. If the second man confirms the first man's assertion, it proves that he has told the truth, that is, he himself really is a Betan. If the third man gives the first the lie, it proves that he is a liar, that is, an Atan. Accordingly the first two are Betans, the last an Atan.

13 — Three will be sufficient because that would allow for four possible combinations: three black ones, three brown ones, a black one and two brown ones, and a brown one and two black ones. All four combinations will allow for one pair of either color, black or brown.

14 — This is the way the smartest of the three — let us call him A — reasoned: "If I had a white spot, B would immediately conclude from the fact that C raised his hand — which in that case could only be attributed to B having a black spot — that he, B himself, was marked with a black spot. Since B didn't reason that way it is impossible that I have a white spot."

15 — If Bill and Cal were wearing red hats, Abe would have known that his hat was blue, because there were only two red hats. Since Abe didn't know the right answer, Cal concluded that there remained only two possibilities for himself and Bill. Either both had blue hats or one had a blue, the other a red hat. If he himself had a red hat, Cal reasoned, Bill would have concluded that he, Bill, had a blue hat, because otherwise Abe would have known that he, Abe, must have a blue hat. So Bill, because he was not able to tell correctly the color of his own hat, involuntarily betrayed to Cal that his hat was not red. Therefore, Cal could tell that his hat was blue.

16 — The problem can only be solved by trial and error, keeping in mind the two stipulations that each witness has made three wrong statements and that each question had been answered correctly at least once. If we assume blond to be the correct hair color, brown, red and black are wrong. Since B and F, like all the witnesses, have answered only one question correctly, the color of the criminal's eyes cannot be black or brown. It must therefore be blue, which gives one correct answer to both A and D. Consequently, grey, dark blue and "not dark brown" are eliminated as the color of the burglar's suit, which must be dark brown. Finally, the criminal's age must be 28 because all other age estimates were made by A, B, C, D and F, who have already one correct answer to their credit. This checks because E's answers to questions I, II and III are wrong and so his answer to question IV must be correct, as stipulated.

By chance, we started with a correct answer (blond for the color of the criminal's hair). If we start with an incorrect statement we will rapidly strike a snag by finding that fulfillment of the two conditions of the problem becomes impossible.

We can, however, start with red for the hair color, and will get another correct solution, namely: eyes, blue; suit, dark blue; age, 28.

17 — These were the shifts of the four watchmen: Jake from midnight to 6 a.m. and from 10 a.m. to 4 p.m., Dick from 4 a.m. to 10 a.m. and from 4 p.m. to 10 p.m., Fred from 10 p.m. to 4 a.m. and from noon to 6 p.m., and Eddie from 6 a.m. to noon and from 6 p.m. to midnight.

18 — The clues allow us to write down the names $(M, S$ and $K)$ of some of the policemen (p) and burglars (b), as follows:

$$p \quad p \quad p \quad p \quad p \quad p \qquad b \quad b \quad b \quad b$$
$$M \; S \; S \; M \qquad\qquad M \; S \; K$$

If we now fill in the names of the last three Millers not yet mentioned — remember, there were six Millers, all told — we will have found out that four policemen were called Miller and two Smith, while two burglars were named Miller, one Smith and one Kelly.

20 — The best way to solve this involved problem is to draw a diagram (Fig. 15 on next page) with "tracks" for each of the four factors involved, that is, first name, last name, hair color and home town, so that each track crosses each of the three others. Where the tracks cross each other we mark with a triangle all direct negative statements, that is, those negative statements which follow directly from the conditions of the problem; we mark with an asterisk all indirect negative statements, that is, those following indirectly from the premises of the puzzle; and with a circle and dot we mark all positive results we reach by eliminating triangle or asterisk possibilities.

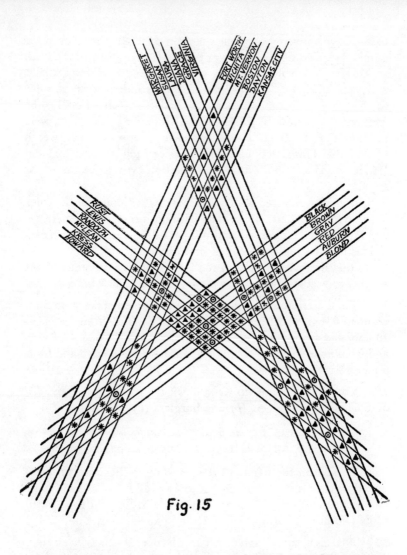

Fig. 15

For instance, using the double-game combinations and the combinations for the first series of singles we will find that Grace cannot be anybody's wife but Rust's. Therefore, in the line formed by the "crossing" of Grace's name with the six family names we insert five triangles and one circle (for Rust). Since Grace is Rust's wife, none of the other women can be married to him, so in the line formed by the crossing of Rust's name with the remaining five first names of the women we insert asterisks.

114

This way we proceed step by step through the whole match, using every bit of information and eliminating all impossibilities. If, for example, we have found out that Grace has brown hair and, later, that the lady with brown hair doesn't hail from Dayton, Ohio, we may eliminate other possibilities for Grace and Rust and mark them with asterisks. The diagram shows only part of the solution, and we leave it to your ingenuity to find the fields in the seeming spider web where the remaining circles belong. This is the complete solution: Laura Howard, gray, from Fort Worth; Diana Kress, blond, from Mt. Vernon; Margaret McLean, red, from Dayton; Virginia Randolph, auburn, from Boston; Susan Lewis, black, from Wichita; and Grace Rust, brown, from Kansas City.

21 — Grelly is a playwright, Pinder is a historian, George writes humorous books, Bird is a poet, Winch writes short stories, and Blank writes novels.

22 — The trip around the mountain need not take longer than 18 days, if division of the day's rations into half rations is permitted. The party itself is not divided. To dump one and one-half rations at 5 miles and return to camp requires half a day. One day is needed for depositing one and one-half rations at 10 miles. To supplement that day's provisions, half a day's ration may be taken from the food stored at 5 miles. Two days are required to get one ration to the 20-mile point. Now we have stores of one-half ration at 5 miles and of one ration each at 10 and 20 miles. Within three more days, one-half ration may be dumped at 30 miles. That adds up to $6\frac{1}{2}$ days. The same number of days is required for dumping half-ration stores at the other end of the circuit, at 90, 80 and 70 miles. Add to these 13 days five more for the final hike around the mountain, and you get 18 days, all told.

23 — Since Reynolds rides the winning horse and Finley's horse is last, two combinations — names of horses and jockeys beginning with the same letter—are out of the question for entry number 2. We can eliminate Reynolds-Riotinto and Finley-Fedor. Only Star can be second, and the name of his jockey may be Shipley, Scranton or Semler. Condor, whose odds equal his position at the finish, cannot be the winner, nor can he have come in third, coincident with the lowest odds, 1 to 1. Neither can he have come in fourth, coincident with the odds 6 to 1. Therefore, Condor is fifth and his odds are 5 to 1. Hence Semler's horse is fourth. Riotinto, whose jockey is not Reynolds, has the same post position and rank at the finish. They cannot be 3 and thus must be 4. Tornado's position at the finish is one higher than his post position number. Thus, it is out of the question that Tornado could have been the winning horse. Therefore, his post position number must have been 2 and he must have come in third at the finish. Because post position and finishing rank are in no case identical, with the exception of Riotinto, the post position number of Reynolds' horse is 5 and that of Finley's horse, Condor, 1. This leaves Reynolds' horse, Fedor, as the winner of the race, with the odds 2 to 1. The odds on Shipley's horse are the same as his post position number, which means that the odds on his horse, Star, are 3 to 1. Thus, Scranton rides Tornado.

This is the complete solution of the problem:

Finish	Jockeys' Names	Horses	Post Positions	Odds
1	Reynolds	Fedor	5	2 to 1
2	Shipley	Star	3	3 to 1
3	Scranton	Tornado	2	1 to 1
4	Semler	Riotinto	4	6 to 1
5	Finley	Condor	1	5 to 1

24 — We can make the following statements and conclusions: (1) Bertha, Celia, Dorothy and Elsa wore the same dresses on the 1st, the 19th and the 31st, that is with intervals of 18 and 12 days. The number of dresses they own, therefore, must be a common divisor of 12 and 18 — that is, any of the numbers 1, 2, 3 and 6. (2) Dorothy has six dresses. She cannot own only one or two because we have already been told that she wore three different colors. On the other hand, she cannot have only the three dresses listed because, in that case, on the 22nd she would have worn the same green dress she wore on the 19th. On the 22nd, however, she wore a yellow one. (3) Emily wore her grey dress on the 1st and the 31st of the month. Consequently, the number of her dresses must be a divisor of 30 — that is, any of the whole numbers 1, 2, 3, 5 and 10. According to (1), 1, 2, 3 and 6 are the numbers pertaining to the other four girls. Therefore, Emily owns either five or ten dresses and consequently, in either of these cases, she had on a grey dress on the 11th. (4) None of the five girls wore a yellow dress on the 11th; therefore, we may be sure that Elsa did not wear a yellow dress on this day and, consequently, owns more

than one dress. That means that she owns either two or three dresses. If she had only two, she would have worn a yellow one on the 11th, just as she did on the first of the month. However, that is impossible since on the 11th nobody wore a yellow dress. So Elsa owns three dresses. Bertha and Celia own one or two dresses, respectively, which means that on the 11th both must have made their appearance in red, one of them because she owns but one dress, the other, who owns two, because on the 11th she must have worn the same dress she wore on the 1st. (5) Dorothy had on a green dress on the 31st and eight days before, on the 23rd, she wore a white dress. That means that she must also have worn a white dress eight days previous to the 19th, because on the 19th she also wore a green dress. So on the 11th she could not have worn a lilac-colored dress either. Consequently, on that crucial day, only Elsa could have worn the lilac-colored dress, since of the five dresses worn that day the four others were worn by Emily, Bertha, Celia and Dorothy.

25 —

```
                          8 4 0 6 3
    1 5 9 2 7 ) 1 3 3 8 8 7 1 4 0 1
              1 2 7 4 1 6
                6 4 7 1 1
                6 3 7 0 8
                  1 0 0 3 4 0
                    9 5 5 6 2
                      4 7 7 8 1
                      4 7 7 8 1
                              0
```

26 —

```
                        2 9 6 5 8
    3 7 0 1 4 ) 1 0 9 7 7 6 1 2 1 2
              7 4 0 2 8
              3 5 7 4 8 1
              3 3 3 1 2 6
                2 4 3 5 5 2
                2 2 2 0 8 4
                  2 1 4 6 8 1
                  1 8 5 0 7 0
                    2 9 6 1 1 2
                    2 9 6 1 1 2
                              0
```

27 —

```
                              4 2 9 7 5 3
         3 5 7 9 2 4 ) 1 5 3 8 1 8 9 1 2 7 7 2
                       1 4 3 1 6 9 6
                       ─────────────
                         1 0 6 4 9 3 1
                           7 1 5 8 4 8
                         ─────────────
                           3 4 9 0 8 3 2
                           3 2 2 1 3 1 6
                           ─────────────
                             2 6 9 5 1 6 7
                             2 5 0 5 4 6 8
                             ─────────────
                               1 8 9 6 9 9 7
                               1 7 8 9 6 2 0
                               ─────────────
                                 1 0 7 3 7 7 2
                                 1 0 7 3 7 7 2
                                 ─────────────
                                             0
```

28 —

```
                  1 4 1 9
     9 4 6 ) 1 3 4 2 3 7 4
             9 4 6
             ─────
             3 9 6 3
             3 7 8 4
             ─────
               1 7 9 7
                 9 4 6
               ─────
                 8 5 1 4
                 8 5 1 4
                 ─────
                       0
```

29 —

```
                    4 5 3 6 7
     2 9 1 8 ) 1 3 2 3 8 0 9 0 6
               1 1 6 7 2
               ─────────
                 1 5 6 6 0
                 1 4 5 9 0
                 ─────────
                   1 0 7 0 9
                     8 7 5 4
                   ─────────
                     1 9 5 5 0
                     1 7 5 0 8
                     ─────────
                       2 0 4 2 6
                       2 0 4 2 6
                       ─────────
                               0
```

30 —

```
                    3 0 0 5 0 2 8
     1 9 9 ) 5 9 8 0 0 0 5 7 2
             5 9 7
             ─────
               1 0 0 0
                 9 9 5
               ─────
                   5 5 7
                   3 9 8
                   ─────
                     1 5 9 2
                     1 5 9 2
                     ─────
                           0
```

This is only one of several possible solutions.

118

```
                    6 5 2
  3 9 2 6 ) 2 5 5 9 7 5 2
            2 3 5 5 6
              2 0 4 1 5
              1 9 6 3 0
                    7 8 5 2
                    7 8 5 2
                          0


                    9 7 8 0 9
  1 2 4 ) 1 2 1 2 8 3 1 6
          1 1 1 6
              9 6 8
              8 6 8
              1 0 0 3
                9 9 2
                  1 1 1 6
                  1 1 1 6
                        0


                  8 0 8 0 9
  1 2 4 ) 1 0 0 2 0 3 1 6
          9 9 2
          1 0 0 3
            9 0 2
              1 1 1 6
              1 1 1 6
                    0


                  2 4 6 7 8
  3 9 5 0 1 ) 9 7 4 8 0 5 6 7 8
              7 9 0 0 2
              1 8 4 7 8 5
              1 5 8 0 0 4
                2 6 7 8 1 6
                2 3 7 0 0 6
                  3 0 8 1 0 7
                  2 7 6 5 0 7
                    3 1 6 0 0 8
                    3 1 6 0 0 8
                            0
```

```
                  1 0 1 1 . 1 0 0 8           7 7 5
6 2 5 ) 6 3 1 9 3 8                         × 3 3
        6 2 5                               2 3 2 5
          6 9 3                           2 3 2 5
          6 2 5                           2 5 5 7 5
            6 8 8
            6 2 5
              6 3 0
              6 2 5
                5 0 0 0
                5 0 0 0
                      0
```

37 — The basic number must be such that its square has seven and its cube has eleven digits. That gives as the limits between which you will have to look for the basic number the numbers 2155 and 3162. (Try 2154 and 3163 and you will see why.) The position of the asterisks in the second multiplication skeleton indicates that the next to the last numeral of the square must be a zero. Now, only numbers ending with 00, 01, 02, 03, 47, 48 and 49 have squares with a zero as the next to the last digit. The skeleton for the first multiplication shows that the basic number has four digits, the first three of which must be below 5, while the last one must be 5 or more. For all these reasons, the basic number must be a four-digit figure beginning with the number pair 22, 23 or 24 and ending with the number pair 47, 48 or 49. Since you haven't much choice left, it won't take you long to find out, by trial and error, that the number fulfilling these conditions can only be 2348.

38 —

$$11664 + \frac{\begin{array}{r}21025\\20736\end{array}}{289} = 32400 \quad \text{equals} \quad 108^2 + \frac{\begin{array}{r}145^2\\144^2\end{array}}{17^2} = 180^2$$

$$18225 + \frac{\begin{array}{r}32761\\32400\end{array}}{361} = 50625 \quad \text{equals} \quad 135^2 + \frac{\begin{array}{r}181^2\\180^2\end{array}}{19^2} = 225^2$$

$$27225 + \frac{\begin{array}{r}48841\\48400\end{array}}{441} = 75625 \quad \text{equals} \quad 165^2 + \frac{\begin{array}{r}221^2\\220^2\end{array}}{21^2} = 275^2$$

$$15876 + \frac{\begin{array}{r}28900\\28224\end{array}}{676} = 44100 \quad \text{equals} \quad 126^2 + \frac{\begin{array}{r}170^2\\168^2\end{array}}{26^2} = 210^2$$

$$28224 + \frac{\begin{array}{r}51076\\50176\end{array}}{900} = 78400 \quad \text{equals} \quad 168^2 + \frac{\begin{array}{r}226^2\\224^2\end{array}}{30^2} = 280^2$$

39 –

```
  6 0 5 4        7 8 9 4
  1 7 2 0        1 0 3 8
  9 7 3 4        2 0 5 4
  ───────        ───────
 1 7 5 0 8      1 0 9 8 6
```

40 –

```
  9 4 8 6        9 3 7 6
  1 0 7 6        1 0 8 6
 ───────        ───────
 1 0 5 6 2      1 0 4 6 2
```

41 –

```
  8 4 3 2
  8 4 7 5
 ───────
 1 6 9 0 7
```

42 –

$$19 + 91 + 9 = 119$$

43 –

$$17 \times 2 = 34 \quad 17 \times 4 = 68$$
$$90 - 34 = 56 \quad 93 - 68 = 25$$

44 –

$$999 \times 999 + 999 = 999000$$

45 –

$$3 \times 6 = 18,$$
$$18 \times 54 = 972.$$

46 –

$$
\begin{array}{ccc}
10 & \times 16 = & 160 \\
+ & + & - \\
14 & \times 8 = & 112 \\
\hline
24 & + 24 = & 48
\end{array}
$$

47 –

$$
\begin{array}{ccc}
10 & \times 26 = & 250 \\
+ & + & - \\
14 & \times 14 = & 196 \\
\hline
24 & + 40 = & 64
\end{array}
$$

48 –

$$
\begin{array}{ccc}
6 & \times 46 = & 276 \\
+ & + & - \\
14 & \times 14 = & 196 \\
\hline
20 & + 60 = & 80
\end{array}
$$

49 –

$$
\begin{array}{ccc}
2 & \times 290 = & 580 \\
+ & + & - \\
16 & \times 16 = & 256 \\
\hline
18 & + 306 = & 324
\end{array}
$$

50 –

$$
\begin{array}{ccc}
6 & \times 81 = & 486 \\
+ & + & - \\
19 & \times 19 = & 361 \\
\hline
25 & + 100 = & 125
\end{array}
$$

51 –

$$
\begin{array}{ccc}
46 & \times 126 = & 5796 \\
+ & + & - \\
74 & \times 74 = & 5476 \\
\hline
120 & + 200 = & 320
\end{array}
$$

52 –

$$
\begin{array}{ccc}
26 & \times 15626 = & 406276 \\
+ & + & - \\
624 & \times 624 = & 389376 \\
\hline
650 & + 16250 = & 16900
\end{array}
$$

53 –

$$
\begin{array}{ccc}
26 & \times 7480226 = & 194485876 \\
+ & + & - \\
13674 & \times 13674 = & 186978276 \\
\hline
13700 & + 7493900 = & 7507600
\end{array}
$$

54 – Eight sons and 56 grandsons. MacDonald was 64. We had to tell you that MacDonald was under 80, because the other conditions of the problem would have been fulfilled if he had 9 sons and 72 grandsons, except for the detail that then he would have to be 81 $(9 + 72)$. We also stipulated under 50 as he could have had 7 sons and, consequently, 42 grandsons, but that would be somewhat excessive for a man of 49!

55 – The ratio will be 1 : 2. If the difference in years between "now" and "once" is d, the present ages of Mr. and Mrs. Wright, m and f respectively, and the present ages of the children a and b, we can write down two equations: $f + a + b = m$ and $a + b + 2d = f + d$ (or $a + b + d = f$). Consequently, $m + d = f + a + b + d = 2f$.

56 – The manager is 36, because, x years ago, when the foreman was as old as the manager is now, the manager was 24 (half as old as the foreman is now). Thus, we have $48 - x = m$; $m - x = 24$. Consequently, $2m = 72$ and $m = 36$.

57 – The ages of the four people are 97, 79, 31 and 13 years. If we designate the four people with x, y, u and v, in that order, we have $x - y = u - v$. Now all four figures must consist of two digits (because you must be able to turn them around – which excludes a one-digit figure for the great-grandson – and three-digit figures obviously wouldn't be common sense. If we choose digit pairs with the difference 1, that is, 12, 23, 34, 45, 56, 67, 78, 89, we don't get any primes which, when turned around, yield more primes. Therefore, we have to choose another difference to pick our four primes. Only when we try the difference 2 will we get four primes which fill the above equation.

58 – 10 years.

59 – 18 years.

60 – Jim is 53 years and 4 months old and his sons are 26 years 8 months, 13 years 4 months, and 6 years 8 months, respectively.

61 – She married when she was 17. Now she is 27, her son is 9 and her daughter 3.

62 – Bill is 40, Sam 30. When Bill was 30, Sam was 20. When Sam is 40, Bill will be 50 and both together will be 90.

63 – 18 years.

64 – Marion is $29\frac{2}{5}$, Betty $19\frac{3}{5}$ years of age.

65 – $10\frac{1}{2}$ and six months.

66 – Let x be one-half the father's present age and a, b, c and d be the ages of the four children, in that order, beginning with the youngest. We have a number of equations: (1) $a + b + c + d = x$, (2) $d^2 - a^2 = 2x$, (3) $d^2 - c^2 + b^2 - a^2 = x$, (4) $a + b + c + d + 64 - 2x = 2x + 16$, or $a + b + c + d + 48 = 4x$. Subtracting (1) from (4) we have $3x = 48$, or $x = 16$. The father's age, therefore, is 32.

From (2) and (3), now substituting 16 for x, we get $c^2 - b^2 = 16$, and from (1) and (3) we get $b = a + 1$ and $d = c + 1$. Now (1) yields $a + c = 7$ and $a + d = 8$. Since we know that $d^2 - a^2 = 32$, we now find $d - a = 4$ and finally $a = 2$, which is the age of the youngest child. Now we have simple equations for the other three unknowns: $b = 3$, $c = 5$, and d, the age of the oldest child, $= 6$.

67 — We designate the four couples with A and a, B and b, C and c, D and d. Then the following crossings are necessary:

Left Shore	River	Island	River	Right Shore
AaBbCcDd	—	—	—	—
ABCcDd	ab→	—	—	—
ABCcDd	←b	a	—	—
ABCDd	bc→	a	—	—
ABCDd	c	ab	—	—
CcDd	AB→	ab	—	—
CcDd	—	AB	ab→	—
CcDd	—	AB	←b	a
CcDd	—	b	AB→	a
CcDd	—	b	←B	Aa
CcDd	←B	b	—	Aa
BCD	cd→	b	—	Aa
BCD	←d	bc	—	Aa
Dd	BC→	bc	—	Aa
Dd	—	bc	BC→	Aa
Dd	—	bc	←a	ABC
Dd	—	c	ab→	ABC
Dd	—	c	←C	AaBb
Dd	←C	c	—	AaBb
d	CD→	c	—	AaBb
d	—	c	CD→	AaBb
d	—	c	←b	ABCDa
d	—	—	bc→	AaBCD
d	—	—	←c	AaBbCD
d	←c	—	—	AaBbCd
—	—	—	cd→	AaBbCD
—	—	—	—	AaBbCdDd

68 — Jones shipped his $5000 worth of nuggets to the other shore and returned, then Smith crossed with only $3000 worth of nuggets. Now $8000 worth of gold remained on the near side of the river. After Smith had come back, all three prospectors were again on the near side of the stream and nuggets valued at $8000 were on each shore. Now Jones and Brown crossed the river and Jones returned with $5000 worth of gold. Then Smith took $8000 worth of gold across and let Brown return with $3000. Then both

Jones and Brown rowed across, leaving $8000 worth of nuggets on the near shore. Smith returned to fetch $5000 worth out of this $8000 and finally Brown came back for the remaining $3000.

69 – Let's call R the cannibal who knows how to row. First, R, in two return trips, rows the other two man-eaters to the other shore and comes back. Then two of the missionaries make the trip and one of them returns with one of the cannibals. Then one of the missionaries crosses with the cannibal, R, and returns with the other cannibal. Now the two remaining missionaries row across so that all three have made the passage safely. Finally, the cannibal, R, gets the two other savages across the river.

70 – If we follow the movements of the two hands beginning with 12, it is easy to see that the two will first meet shortly after 1 o'clock. Since the minute-hand moves 12 times as fast as the hour-hand, the minute-hand will pass 12 times as many minute markings as the hour-hand will, during the time the two hands move from their 1-o'clock positions to the point where they meet. If x is the distance covered on the dial by the hour-hand we get the equation $12x = 5 + x$, which yields $x = \frac{5}{11}$. This is the position on the dial of the hour-hand after 1, and that means that both hands are at $1:5\frac{5}{11}$ when their first meeting takes place. After another 1 hour and $5\frac{5}{11}$ minutes, the two will meet again. Altogether, within 12 hours, this event will take place 11 times, the eleventh time at 12 o'clock.

For the solution of many clock puzzles one may use the following general relations. If the hour-hand is at x (of the minute dial) and the minute-hand at y, $y = 12x - 60k$, with k representing one of the numbers 0, 1, 2, 3 .... 11.

To solve the second part of the problem you use this general equation and the special equation, $x - y = \pm 15$. Hence, $-11x + 60k = \pm 15$, and $x = \dfrac{60k - 15}{11}$ ($k = 1, 2 \ldots 11$), and $x = \dfrac{60k + 15}{11}$ ($k = 0, 1 \ldots 10$). Altogether the two hands will be at right angles 22 times within 12 hours, the first time when the hour-hand points at $1\frac{4}{11}$ minutes and the minute-hand points at $12 \times 1\frac{4}{11} = 16\frac{4}{11}$, that is, $16\frac{4}{11}$ minutes after 12 o'clock.

71 – We have the equation $x - y = 2$. Using again our general equation we get $x - 2 = 12x - 60k$; $11x - 60k + 2 = 0$. x and k must be integers. This Diophantine equation has the only possible solution $x = 38$, $k = 7$. The time in question, therefore, is 7:36.

However, the equation $y - x = 2$ will also solve the problem. Using the general equation $y = 12x - 60k$, we get the Diophantine equation $11x - 60k - 2 = 0$, with the only possible solution $x = 22, 6 = 24$, yielding the time 4:24.

72 – The problem states that first the hour-hand is at x and the minute-hand at y and after reversing them the hour-hand is at y and the minute-hand at x. According to the general formula, x, that is, the position of the minute-hand after the reversal, equals $12y - 60k$. For the first position we have the formula $y = 12x - 60l$. Consequently, $143x = 60 \ (k + 12l)$ or $143x = 60m \ (m = 0, 1, 2 \ldots 142)$. Therefore, exchangeability occurs 143 times every 12 hours.

For example, take $m = 44$. Hence, $x = 2640; 143 = 18\frac{6}{13}$. $y = 221\frac{7}{13} - 180 = 41\frac{7}{13}$, that is, $3:41\frac{7}{13}$ o'clock. Reversal of the hands yields $8:18\frac{6}{13}$ o'clock.

73 – Doubtlessly, you will have reasoned, within any one minute, the second-hand will twice be parallel to each of the two other hands. Thus, you may have figured there are 5,760 such parallel positions in one day. However, in that case, you overlooked one little thing. The hour-hand and the minute-hand together make 26 revolutions in 24 hours. Therefore 26 parallel positions have to be deducted. Consequently, the correct figure is 5,734.

74 – The positions of the three hands are x, y and z. Then we have first, $y - x = 20$ or $11x - 60k = 20$ and $x = \dfrac{60k + 20}{11}$, and second, $z - x = 40$ or $719x - 60l = 40$ and $x = \dfrac{60l + 40}{719}$. Consequently, $\dfrac{3k + l}{11} = \dfrac{3l + 2}{719}$ or $3 \ (11l - 719k) - 697$. This Diophantine equation has no integral solutions, which means that the points of the three hands will never form an equilateral triangle. One may find the triangle closest by replacing 697 by 696. Then $11l - 719k = 232$ and $k = 8, l = 544$, and $x = 45\frac{5}{11}, y = 5\frac{5}{11}, z = 27\frac{3}{11}$, which corresponds to $9:5:27\frac{3}{11}$ o'clock. If we change the order of the three hands, for instance, by exchanging x and y, we can similarly prove the impossibility of an equilateral triangular position.

75 – In order to show the same time, one clock must have lost 12 hours while the other gained that much. Two of the clocks losing and gaining, respectively, one minute every day, this moment must arrive in 720 days. Now the computation is simple unless you forgot that 1900 was not a leap year. The expected

125

event will have occurred on March 22, 1900.

76 — Within 12 hours, one or the other of the two keyholes is covered for $2 \times 3 \times 12 = 72$ minutes by the minute-hand and for $2 \times 36 = 72$ minutes by the hour-hand. From these figures 6 minutes have to be deducted because both hands cover the same hole simultaneously. Therefore, within 12 hours, both keyholes will be uncovered for $720 - 72 - 72 + 6 = 582$ minutes $= 9$ hours 42 minutes.

77 — Within one full day the clock gains $\frac{1}{6}$ minute. Therefore, if you are a little rash you may have figured that $30 \times \frac{1}{6} = 5$, that is, the 5-minute gain will have been reached after 30 days or at dawn of May 31st. You may, though, have made the same mistake as when you figured how long it would take the snail to get out of the well (problem 6). As a matter of fact, the clock gained 5 minutes some time earlier. On the 28th, it has gained $27 \times \frac{1}{6} = 4\frac{1}{2}$ minutes. On the following day, the 29th, it will gain another half minute and thus be 5 minutes fast.

78 — If the positions of the long and short hands are x and y, in the reflected image they are $x' = 60 - x$, and $y' = 60 - y$. Now $y' = 60 - y = 60 - 12x + 60k = 12 (5 - x) + 60k = 12 (60 - x) + 60k - 660 = 12x' - 60 (11 - k)$. Since $k = 0, 1, 2 \ldots 11$, the same relation stands for $k' = 11 - k$. Consequently, $y' = 12x' - 60k'$ $(k' = 0, 1, 2 \ldots 11)$. It follows that the reflection of a clock always shows a correct time reading.

79 — The problem can be solved by starting with the youngest child who, according to the provisions of the will, has to receive so many thousand dollars outright. Then the share of the next to the last child has to be computed and so on until all the shares have been figured out. If, for instance, there are 5 children, the youngest will receive $5,000; the next oldest, $4,000 + 2,500 = $6,500; the next, $8,750 [or, $3,000 + \frac{1}{2} (6,500 + 5,000)$]; the next, $12,125; and the eldest, $17,187.50. That adds up to $49,562.50. But the fortune consisted of a full amount of dollars and therefore neither of the two wills could have been drawn up when the testator had five children. For the same reason, it is easy to prove that all solutions except those for four and six children are no good. When the testator had four children he owned $24,250 and drew up his first will. When he had six children he owned $95,125 and drew up his second testament, which was, in principle, the same as the first. In the first case, the shares of the four children were, starting with the youngest, $4,000; $3,000 + \dfrac{4,000}{2} = $5,000;

126

$2,000 + \dfrac{9,000}{2} = \$6,500$; and $1,000 + \dfrac{15,500}{2} = \$8,750$; altogether $\$24,250$. In the second case, the shares of the six children were $\$6,000$; $5,000 + \dfrac{6,000}{2} = \$8,000$; $4,000 + \dfrac{14,000}{2} = \$11,000$; $3,000 + \dfrac{25,000}{2} = \$15,500$; $2,000 + \dfrac{40,500}{2} = \$22,250$; $1,000 + \dfrac{62,750}{2} = \$32,375$; altogether $\$95,125$.

The general solution of the problem is as follows: If n is the number of children, the estate $S = 1,000 \ (n\text{-}2) \dfrac{3^n}{2^{n}\text{-}1} + 4,000$. The shares of the children are $A_k = 1,000 \ (n\text{-}2)\left(\dfrac{3}{2}\right)^{n-k} + 2,000$. ($k = 1, 2, 3 \ldots \ldots n$). S and A_k yield full dollar amounts for $n = 1, 2, 3, 4$ and 6. In this case, only $n = 4$ and $n = 6$, with $S = 24,250$ and $95,125$, respectively, can be used. (You will remember the testator had two more children when he drew up his second will.)

80 — If x, y and z are the numbers of full, half-full and empty casks which each of the sons gets, and if the contents of a full cask are arbitrarily set at 2 and that of a half-filled cask at 1, you can write down the two equations $x + y + z = 8$ and $2x + y = 7$, which have the following solutions:

x	y	z
0	7	1
1	5	2
2	3	3
3	1	4

This shows that there are four alternatives for letting a son get his share in the estate, three of which have to be combined in such a way that the sum of the three x values amounts to 5. This may be done in three different ways: a) $0 + 2 + 3$; b) $1 + 1 + 3$; c) $1 + 2 + 2$. Consequently, the three sons, A, B and C, may distribute the heritage in any of the three following ways:

	x	y	z		x	y	z		x	y	z
A	0	7	1	A	1	5	2	A	1	5	2
B	2	3	3	B	1	5	2	B	2	3	3
C	3	1	4	C	3	1	4	C	2	3	3

81 — As in the previous problem you have to find the equations controlling the conditions of the problem, $x + y + z + u =$

$\frac{228}{19} = 12$ and $3x + 2y + z = \frac{456}{19} = 24$. These equations yield the following series of results:

z	x	y	u
0	0	12	0
0	2	9	1
0	4	6	2
0	6	3	3
0	8	0	4
1	1	10	0
1	3	7	1
1	5	4	2
1	7	1	3
2	2	8	0
2	4	5	1
2	6	2	2
3	3	6	0
3	5	3	1
3	7	0	2
4	4	4	0
4	6	1	1
5	5	2	0
6	6	0	0

The table shows that there are 19 different alternatives for letting a person share in the heritage according to the stipulations of the will. Since the club had 19 members, these 19 possibilities represent the only solution of the problem.

82 — Each dress cost $\frac{\$1800.}{x}$ If he had bought them a day earlier the store owner would have received $x + 30$ dresses for $1800. Consequently, each dress would have cost him $\frac{\$1800.}{x+30}$ Since each dress would have been $3 cheaper than the price he eventually paid, you can write down the equation $\frac{1800}{x+30} + 3 = \frac{1800.}{x}$ The positive root of this quadratic equation is 120. Thus, 120 dresses had been bought and 20 of them had been stolen.

83 — The women's names are Jean Smith, Kate March and May Hughes and their shares are $1320, $1220 and $1420, while the husbands inherit, in the same order, $1980, $1220 and $2840.

84 — If we call the length of the bridge $3d$, the two vehicles will meet when the car has travelled $2d$ and the truck d. To reach this point the car has required t seconds and the truck $2t$ seconds. To back up, the car would need $2t$ seconds for the distance $2d$, the

truck $4t$ seconds for the distance d. Now let's consider both possibilities:

Case 1. The car backs up. It requires $2t$ seconds. Meanwhile, the truck proceeds. It needs $4t$ seconds to finish the journey across the bridge so the car has to wait $2t$ seconds until it can start across again. It requires $1\frac{1}{2}t$ seconds to drive over the bridge. Thus, altogether, $5\frac{1}{2}t$ seconds are needed for the operation.

Case 2. The truck backs up. It requires $4t$ seconds. Simultaneously, the car proceeds, though only with one-eighth of its speed because of the truck's snail's pace. They arrive at the end of the bridge together. Then the truck drives over the bridge, which takes $6t$ seconds. Thus, in this case, 10 seconds altogether are needed for the operation. Therefore, Case 1 is the solution most favorable for both vehicles.

85 — The length of the trunk is x. While the pedestrian moves one yard, the vehicle moves y yards. Thus, while the hiker moves 140 yards, the tip of the tree has moved $140y$ yards forward. Therefore, 140 yards equal $x + 140y$ yards. When the pedestrian moves in the other direction it takes him only 20 paces to pass the tree while, meanwhile, the tip of the tree has moved $20y$ paces, or yards forward. Both distances add up to the length of the trunk. Therefore, we have two equations: $x + 140y = 140$ and $20 + 20y = x$. $x = 35$ yards.

86 — The pedestrian meets 20 cars an hour and 10 cars an hour pass him. Thus, within one hour, 30 cars altogether make the run, 15 in each direction. Therefore, every four minutes a car leaves each terminal.

87 — The number of visible steps, x, is the sum of those counted by either of the boys, 50 and 75, respectively, plus those the escalator moved while the boy in question was climbing it. The two parts of the sum have the same ratio as the speed of the boys, s and $3s$ respectively, and that of the escalator, s_e. Therefore, $s : s_e = 50 : (x - 50)$ and $3s : s_e = 75 : (x - 75)$; $150 (x - 75) = 75 (x - 50)$; $x = 100$ steps. By the way, the little boy's speed is exactly the same as the speed of the escalator.

88 — Simple reasoning will indicate that the first bicyclist will have to drop the vehicle at some point after having passed the summit of the hill. During his trip all three speeds occur, while in the case of the second boy only single and triple speeds need be considered. Using again the equation $t = \dfrac{d}{s}$ you can quickly find the answer to the question. The first boy has to drop the bicycle after having bicycled $\frac{9}{16}$ of the route. The equation reads, where a is the boy:

$$\frac{\frac{a}{2}}{2s}+\frac{a\left(x-\frac{1}{2}\right)}{3s}+\frac{a\left(1-x\right)}{s}=\frac{a}{2s}+\frac{a\left(x-\frac{1}{2}\right)}{s}+\frac{a.\left(1-x\right)}{3s}\;;\;x=\frac{9}{16}.$$

89 — The distance, x, the second man has to walk until he meets the motorcyclist, and the distance the first man must walk from the moment he has dismounted until he reaches the station, must be equal because both men are supposed to arrive simultaneously. Therefore, they must have walked and ridden equal lengths of time. The motorcyclist altogether makes $9 - x + 9 - 2x + 9 - x = 27 - 4x$ miles at 18 miles an hour, while each of his two friends, in the same time, makes $9 - x$ miles with a speed of 18 miles an hour, and x miles with a speed of 3 miles an hour. So you have the equation $\frac{27 - 4x}{18} = \frac{9 - x}{18} + \frac{x}{3}$, which yields $x = 2$. Therefore, the first man has to dismount after a ride of 7 miles. The motorcyclist returns and meets the second man at a point 2 miles from the start. The whole journey will take 1 hour and $3\frac{1}{3}$ minutes.

90 — The equation $s = \frac{d}{t}$, applied to both cases, yields $\frac{1}{3} = S + s$ and $\frac{1}{4} = S - s$, where S is the bicyclist's speed on a windless day and s is the wind velocity, both in miles per minute. Consequently, S is $\frac{7}{24}$, which means that on a windless day he can ride one mile in $3\frac{3}{7}$ minutes or 3 minutes 26 seconds. The wind velocity s, by the way, is $2\frac{1}{2}$ miles per hour. With the wind at his back the bicyclist can make 20 miles per hour; against the wind, 15 miles; and on a calm day, $17\frac{1}{2}$ miles.

91 — If S and s are the speeds of the two trains, x the time interval between their meeting and their arrival, and a and b the distances both trains have traveled until they meet, we may write down the equations $a = 80S$, $b = 125s$, $a = sx$ and $b = sx$. We find $x = \frac{80S}{s} = \frac{125s}{S}$. Consequently, $x^2 = 10,000$ and $x = 100$ minutes. The trains arrive at 12:15.

92 — If the whole distance is divided into two parts, e and f, by the meeting point of the trains, the ratio of e and f is the same as that of the speeds S and s, or $9s$ and $4S$. Consequently, the ratio of the speeds is $3 : 2$.

93 – By methods similar to those used in previous problems, we obtain a number of equations which yield 200 miles as the distance between the two towns, 4 hours as the length of the trip according to the timetable and 50 miles per hour as the train's average timetable speed.

94 – Let's call the nine men A, B, C, D, E, F, G, H and I. First, all drive 40 miles on the gas in their tanks. Then A transfers one gallon to each of the eight cars, which leaves him one gallon for his return trip. After the remaining cars have traveled another 40 miles, B transfers one gallon to each of his seven co-explorers and returns on the two gallons left to him. Proceeding this way finally only I remains, to whom H has given one gallon. He travels another 40 miles and returns on his reserve of 9 gallons after having reached a point 360 miles from the desert's eastern edge.

95 – The whole distance is 3 miles. The first runner races $1\frac{1}{2} + \frac{1}{2} = 2$ miles, the second, $\frac{1}{3} + \frac{1}{3} = \frac{2}{3}$ miles and the third, $\frac{1}{12} + \frac{1}{4} = \frac{1}{3}$ mile.

96 – The distance is $1.5t_1$ or $4.5t_2$. The sum of the times, $t_1 + t_2$, is 6 hours. So we have three equations for three unknown quantities and it is easy to figure out the distance. It is $6\frac{3}{4}$ miles.

97 – If we call the distance between Ashton and Beale x, and that between Beale and Carter y, we have the following equations: $25 = y - 5 + \frac{x}{2} = \frac{x}{2} + 10$. The three sides of the triangle measure 20, 15 and 30 miles. The speeds of the two boats are in the ratio of 40 to 25, or 8 to 5.

98 – If x is the distance the head of the army advanced while the dispatch rider rode from the rear to the head of the army, we have the equation $(50 + x) : x = x : (50 - x)$, which gives $x = 35.35$ miles. The dispatch rider rode altogether $50 + x + 50 - (50 - x) = 2x + 50$ miles, which is 120.7 miles.

99 – Bob reaches Al in one hour $(1\frac{1}{2}x + 1\frac{1}{2}x = 3x)$. The dog has run all the time. Since he moves with a speed of 6 miles an hour he has covered exactly 6 miles.

100 – If the length of each candle is 1, and t is the length of time the current was off, the basic formula $d = s \times t$ and the conditions set forth in the problems yield the equations $S = \frac{1}{4}$, $s = \frac{1}{5}$; $1 - x = \frac{t}{4}$, $1 - 4x = \frac{t}{5}$. Consequently, $t = 3\frac{3}{4}$ hours.

101 — If the dinner lasted t hours and we call the lengths of the candle stumps x, $x + y$, $x + 2y$, we can write down the following equations:

$$\frac{1}{1-x} = \frac{4.5}{t}\,; \quad \frac{2}{2-x-y} = \frac{4}{t}\,; \quad \frac{1}{1-x-2y} = \frac{9}{t}\cdot \quad \text{Hence } t = 3 \text{ hours.}$$

102 — After train E has moved the necessary distance to the east, engine E (by itself) moves on to the siding and the three-car train W moves east, past the siding. Engine E returns to the main track to haul three W cars to the west. Then the W engine moves on to the siding. The E locomotive with three cars moves east and then pulls all seven cars westward. Now engine W returns to the main track, pulls five cars beyond the siding and pushes the last car on to the siding and the remaining four westward. Then it returns and, pulling the car standing on the siding onto the main track, couples it to the others. Engine W now pulls six cars to the east, shunts the last onto the siding and pushes five westward. Then engine W again moves eastward with one car, pulls the other car from the siding and then pushes the two cars westward. It then pulls all seven cars to the east, past the siding. The last car is then pushed onto the siding and the other six are pushed westward. Finally, engine E can be coupled to its four cars and continue its trip while engine W pulls the last car from the siding and continues the journey with its three cars.

103 — The local is divided in two. The rear part is shunted onto the siding while the fore part proceeds to the right. Now the express moves forward, then backward onto the siding where the local cars are attached to it and are pulled out and pushed to the left. Now the fore part of the local is shunted onto the side-track. Then the express, with half the local cars, moves beyond the switch and releases the local cars.

104 — Locomotive L pulls car A on to track 1 of the turntable and leaves it there. L moves on to track 6, which is then turned toward the main track. Car B is pulled on to 6, L moves on to 3, which is turned toward the main track. C is pulled on to 3, L moves on to 8, which is turned toward the track. D is pulled on to 8, L moves on to 5, which is turned toward the track. E is pulled on to 5, L moves on to 2, which is turned toward the main track. F is pulled on to 2, L moves on to 7, which is turned toward the track. G is moved on to 7. Then 4 is turned toward the main track and L moves over 4, to 1, and pulls A back on to the main track. 6 is turned toward the main track, L pushes A toward B on 6 and then pulls both *A* and *B* out on the main track. Then 8 is turned toward the main track, *L* pushes *A*, *B* and *C* toward *D* and *E* on 8 and 5 and pulls all 5 cars out on to the main track. Finally, 2 is turned toward the main track and *L* pulls *F* and *G* out from 2 and 7.

105 – *L* pushes *B* on to *a*, returns to *M* over II and then moves on to 1, pushes *A* toward *B* and couples both together, pulls both on to *M* and releases *B*. With *A* it returns to I and pushes *A* on to *a*. Now *L* fetches *B* from *M* and moves it on to I. Then *L* returns to *M*, moves on to II and pulls *A* from *a* on to II.

106 – *L* pushes *B* on to the turntable and moves on over *n* - *m* to I, then pushes *A* toward *B* and moves both cars over the main track in the direction of *n*, where it leaves *B*. It then guides *A* over I on to the turntable, then returns and pulls *A* on to II. *L* then hauls *B* from *n*, moves it in the direction of *m* and pushes it on to I. Finally, *L* returns to *m* - *n*.

107 – The engineer moves far enough in the direction *A* so that the 11th car stands near the switch *T*. Then he detaches the rear cars of the train from car 12 on, moves in the direction of *A*, detaches cars 10 and 11 and pushes them on to the connecting track *TU*. Then he returns to *AB* and, detaching the 12th car from the rear section of the train, attaches it to the fore part and moves with it on to *TU*, pushing the cars 10 and 11 on to the track *UD*. With the cars 9 and 12 now at the rear of the train he then maneuvers the entire coupled section over *VW* to *S*, where he leaves 9 and 12. On his return trip he picks up the cars left on *UD* and *TB*.

108 – Train *A* moves from *a* on to *d*, leaving four cars on the siding. It then moves on over *b* and backs up on the main track toward *a*, far enough to leave room for train *B* on this section of the track. *B* now moves from *b* over *c* toward *a*, then backward over *d* toward *b*, pushing the four *A* cars along to *b*. Now *A* moves on to *d*, letting *B* pass on *c*. *B* leaves the four *A* cars on *c*. *A* now moves on toward *b* and backs up on *c* to have its four rear cars attached.

109 – If we call the amount of grass that grows on one acre in one week one unit, then (1) three cows require in two weeks the grass standing on two acres plus four units; (2) two cows require in four weeks the grass standing on two acres plus eight units; and we can deduce, in accordance with (1), that (3) three cows require in four weeks the grass standing on four acres plus eight units. If we subtract (2) from (3) and divide by four we find that one cow in one week requires the grass standing on half an acre. From this result and from a modification of (2) we can deduce that four units equal the grass standing on one acre. We may conclude from (3) that one cow in six weeks requires the grass standing on three acres. In the same period six acres yield 36 units or as much grass as stands on 9 acres. Therefore, the grass that stands on 6 acres plus that which grows there in six weeks corresponds to the grass standing on 15 acres of pasture, which, considering (3), is sufficient to feed five cows for six weeks.

110 – If we denote the grass standing on the pasture by W, the daily growth by u and the amounts of grass the cow, the goat and the goose consume daily by a, b and c, respectively, in that order, we have the following equations: (1) $a + b = c$, (2) $45c - 45b = W - 45u$, (3) $60c + 60a = W + 60u$ and (4) $90b + 90a = W + 90u$. If we multiply (1) by 90 and subtract it from (4) we get $c = \dfrac{W}{90} + u$. Combining this result with (3) we get $a = \dfrac{W}{180}$ and finally, $b = \dfrac{W}{180} + u$. This yields $W = 180u$ and therefore, $b = \dfrac{W}{90}$ and $c = \dfrac{W}{60}$. The question asked in the problem can be expressed by the equation $x(a + b + c) = W + xu$, which yields the result $x = 36$ days, that is, within 36 days all three animals together would have eaten all the grass standing on the pasture plus that grown in the meantime.

111 – Seventeen eggs. If she bought only sixteen it might happen that there would be only four eggs of each variety. So one extra egg must be added to be sure that at least one variety is represented by five eggs. If V is the number of varieties and C the number of chicks wanted of one variety, the general solution of this problem can be expressed by the formula $V(C - 1) + 1$. (You might have solved problem 13 about the air raid and the socks with the aid of this formula but that case was so much simpler that the result was easier to obtain by a little reasoning than by setting up and using a formula.)

112 – The question boils down to the problem of covering the field to best advantage by regular polygons, the sides of which correspond to the required minimum planting distance, a. Only equilateral triangles, squares and regular hexagons can cover a plane without gaps. A hexagonal pattern will not permit full utilization of the soil unless a plant is also placed in the center of each hexagon. This would divide each hexagon into six equilateral triangles and would thus dispose of the hexagon. So only equilateral triangles and squares remain as alternatives. It is easy to reason why the triangle allows for a better utilization of the soil.

Imagine a line drawn across the field and divided into sections equaling a. Since the field is very large, the line is long and it is divided into a great many sections. Another line parallel to the first, and at the distance a from it, is similarly divided. The two lines yield a strip across the field consisting of squares with the side a. Imagine plants standing at all corners of the squares. Now imagine this strip as a long ladder with rungs only loosely attached to the uprights. If you were to pull in the direction indicated by the arrow (Fig. 16), the corner A would always be a inches away

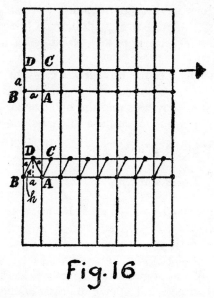

Fig. 16

from B and C, but D would come closer to A until finally it, too, would be only a inches from A. You have now planted as many plants as before, but on a narrower strip of equal length. (The fact that there is one plant less on the upper line may be neglected because, as stated in the problem, the field is very large.) But you have deformed the original strip consisting of squares into one consisting of regular triangles. While the altitude of the original strip

135

is a, that of the new one is the altitude, h, of a regular triangle with the side a, or $\dfrac{a \sqrt{3}}{2}$. In neither case are any of the plants closer to each other than the minimum distance. To find the ratio of the number of plants in each case we have to figure the ratio of the width of both strips, in other words, to divide a by $\dfrac{a \sqrt{3}}{2}$. The result is $2 : \sqrt{3}$ or approximately 1.15. Thus, by dividing the field into regular triangles instead of squares the farmer is in a position to plant about 15 percent more — and to harvest accordingly.

113 — The unit price of the farmer's geese was $\frac{1}{3}$ dollar, that of his neighbor's geese, $\frac{1}{2}$ dollar, hence the average price of each goose of the combined flock should have been $\frac{25}{60}$ dollar.
$\left((\dfrac{30}{3} + \dfrac{30}{2}) \div 60 = \dfrac{25}{60} \right).$ When selling five geese for two dollars, the farmer in effect sold at a unit price of $\frac{2}{5}$ or $\frac{24}{60}$ dollar. Therefore, his unit price was $\frac{1}{60}$ dollar too low and the result, considering that 60 geese were sold, was a deficit of $60 \times \frac{1}{60} = 1$ dollar.

114 — If the younger of the brothers, B, can split only $\frac{3}{4}$ of what the older boy, A, can saw, then A's performance is $\frac{4}{3}$ that of B's. When the younger boy saws and the older one uses the axe, we have the relation $1 : 2$, so we have this tabulation:

While A splits x, $\qquad$ B saws $\dfrac{x}{2}$

While B splits y, $\qquad$ A saws $\dfrac{4y}{3}$

Now $x + y = 5$ and $\dfrac{x}{2} + \dfrac{4y}{3} = 5$. Hence $x = 2$ and $y = 3$. That means that $\quad$ A splits 2, $\qquad$ B saws 1 cord
$\qquad\qquad$ and B splits 3, $\qquad$ A saws 4 cords
$\qquad\qquad$ and thus both split 5 and saw 5 cords.

115 — If we call the original number of chickens x, and of the ducks, y, and the increase factor of the chicken and ducks $3a$ and a, respectively, we have the equation $x + 3ax + y + ay = 200$. We also have the equation $x + y = 25$ for the original flock. Knowing that all unknowns, x, y and a, must be whole numbers, we can determine their values. The farmer's wife started her business with 5 chickens and 20 ducks. She had a 15 fold increase in chickens and a 5 fold increase in ducks. Today she owns 80 chickens and 120 ducks.

116 — 1, 2, 3, 4 . . . . years after the birth of the farmer's cow there will be 0, 1, 1, 2, 3, 5, 8, 13, 21, 34. . . .calves. Each figure in this series is the sum of the two preceding it and the 25th figure is 46,368. If the farmer adds all the 25 figures together he gets 121,392, which is the number of calves the farmer will have had in 25 years.

This series is quite a peculiar one. If you form a sequence of fractions with the members of this set by taking two consecutive members as the numerator and denominator, respectively, you have the sequence, $\frac{0}{1}$ $\frac{1}{1}$ $\frac{1}{2}$ $\frac{2}{3}$ $\frac{3}{5}$ $\frac{5}{8}$ $\frac{8}{13}$ $\frac{13}{21}$ etc., the members tending toward the limit $\frac{\sqrt{5}-1}{2}$.

This value is the length of the side of a regular decagon inscribed in a circle with the radius 1. The Greek mathematicians called the ratio $1 : \frac{\sqrt{5}-1}{2}$, the golden ratio because they believed that a rectangle, the sides of which have this ratio, that is, approximately 1.62, was aesthetically the most satisfactory. Therefore, the figures 0, 1, 1, 2, 3, 5, 8, 13, 21. . . . are sometimes called the golden numbers.

117 — A rope 14½ yards long is pulled through the rings and its ends fastened to the goats' collars. However and wherever the bellicose goats move and graze they can never be closer to each other than ½ yard from collar to collar. At the same time, they are able to graze the entire meadow, including the four corners.

118 — Sullivan was wrong; his damage amounted to only $20. He paid out to the customer the value of one pair of shoes, that is, $12, and $8 change, and to McMahon, $20, altogether $40; but he also received $20 in change from McMahon. Therefore, his loss was only $20.

119 — Mr. Marinacci got only nine different weights for any combination of his five cheeses. It ought to be ten because five different quantities can be combined in ten different ways. Therefore, two different combinations must have resulted in the same weight. If each cheese is combined with each of the others in turn, each will be weighed four times. Hence the sum of the weights of all ten combinations must be divisible by four.

The sum of all nine weights mentioned is 322 pounds. To this figure the missing tenth weight must be added to get four times the weight of all five cheeses. This means that among the weights given in the problem one has to be found which, added to 322, yields a number divisible by four. The only possible figure is 30.

$(322 + 30) \div 4 = 88$, and that is the sum of the weights of all five cheeses. The weights listed in the problem indicate that the two lightest cheeses together weighed 20 pounds and the two heaviest weighed 51 pounds. The remaining fifth cheese, therefore, weighed $88 - (20 + 51) = 17$ pounds. By trial and error you will find in a jiffy that the four remaining cheeses weighed 7, 13, 23 and 28 pounds.

120 — If we denote the number of pairs of socks of each color by n, Mr. Collins paid $3.15n altogether. If we were to use one third of this amount for buying socks of each color and were then to find that he had bought one more pair of socks, we can write the following equation:

$$\frac{3.15n}{3 \times 1.00} + \frac{3.15n}{3 \times 1.05} + \frac{3.15n}{3 \times 1.10} = 3x + 1.$$

From this equation we find n to be 220 and the amount spent to buy $3n = 660$ pairs of socks to be $693. In the second case, Mr. Collins could have bought 231 pairs of black, 220 pairs of brown and 210 pairs of white socks, altogether 661 pairs, one pair more than he really bought.

121 — In a set of squares we write down the amounts any one married couple can have spent.

	I	II	III	IV
May (10)	20	30	40	50
Hull (20)	40	60	80	100
Shelton (30)	60	90	120	150
Ziff (40)	80	120	160	200

Choosing one of the four figures in each column we now have to find a combination which adds up to 400 (500 less total spent by wives). The only such combination is the sum of 20 in the first column, 60 in the second, 120 and 200 in the third and fourth. Thus we find that the names of the four men are Peter May, Henry Hull, Anthony Shelton and Robert Ziff.

122 — With the aid of the simple equation $100x + 2y = 2(100y + x)$ we find that James once had $99.98.

123 — Since all three brands were represented in every purchase, the number of cigars he bought fluctuated between 21 and 27. If we denote the packages by Roman figures, according to the number of cigars they contain, we have the following combinations: 21 cigars—4 III, 1 IV, 1 V; 22 cigars—3 III, 2 IV, 1 V; 23 cigars— 2 III, 3 IV, 1 V or 3 III, 1 IV, 2 V; 24 cigars—2 III, 2 IV, 2 V or 1 III, 4 IV, 1 V; 25 cigars—1 III, 3 IV, 2 V or 2 III, 1 IV, 3 V;

26 cigars—1 III, 2 IV, 3 V; 27 cigars—1 III, 1 IV, 4 V. This tabulation shows that for 23, 24 and 25 cigars the smoker could get two possible combinations.

124 – After having added up the face value of the set, it is easy to figure that each customer will have to get five stamps with a combined face value of 30. This condition can be complied with by two combinations of five different face values and 12 combinations of four different face values. To make the distribution of values as wide as possible, the first group of two combinations is used, supplemented by six combinations picked from the second group.

For instance, the following distribution complies with the conditions of the problem: Customer I gets one each of the face values 15, 5 and 2 and two of the value 4; customer II gets one each of the face values 15, 10 and 1 and two of the value 2; III gets one each of the values 15, 5 and 4 and two of the value 3; IV gets one each of the values 15, 8, 4, 3 and 1; V gets one each of 15, 8 and 5 and two of 1; VI gets one each of 5, 3 and 2 and two of 10; VI gets one each of 10, 8, 5, 4 and 3; and VIII gets one each of 10, 3 and 1 and two of 8.

Face Value	15		10		8		5		4		3		2		1		Total		
Customer	No.	Price	No.	Pr.	No.	Pr.	No.	Pr.	No.	Pr.	No.	Pr.	No.	Pr.	No.	Pr.	F.V.	Pr.	No.
I	1	16					1	6	2	10			1	3			30	35	5
II	1	16	1	11									2	6	1	2	30	35	5
III	1	16					1	6	1	5	2	8					30	35	5
IV	1	16			1	9			1	5	1	3			1	2	30	35	5
V	1	16			1	9	1	6							2	4	30	35	5
VI			2	22			1	6			1	4	1	3			30	35	5
VII			1	11	1	9	1	6	1	5	1	4					30	35	5
VIII			1	11	2	18					1	4			1	2	30	35	5
Total	5		5		5		5		5		5		5		5				40

Fig. 17

Now for the problem of charging the customers equitably. Only the price of the stamp with the face value 2 is fixed at $3. The prices for the other stamps of the set have to be fixed in such a way that the amount charged to each customer is the same. Well, the price of the 2-stamp is $3, that is, one more than its face value. If the prices for the other stamps are figured the same way, each customer will be charged $35. Therefore, the prices for the stamps of the sets ought to be fixed at $16, 11, 9, 6, 5, 4, 3 and 2. With the aid of Fig. 17 you may check whether all the conditions of the problem are met.

125 — Each got $40 back in his stamp. Sheridan had invested $50 and so his share in the third stamp was $10. Malone had invested $70 and so his share in it was $30. Hence they would have to distribute the proceeds of their sale at the ratio of 10 : 30. That means that out of their profit Sheridan got $30 and Malone $90.

126 — Measure out on the disk an arc of 13 inches ($a - f$, Fig. 18)

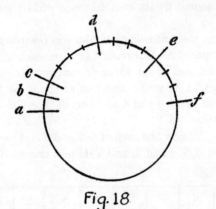

Fig. 18

and then mark on this arc, starting at a, four more points, b, c, d and e, at the distances 1, 1, 4 and 4 inches. You may try out whether indeed any length between 1 and 27 inches can now be measured out on the disk ($e - f$ is 3 inches). However, this is not the only possible solution of the problem. Others are:

1	1	3	8	9	5	1	2	4	5	8	7	1	2	13	4	1	6	1	4	2	8	3	9
1	1	4	3	7	11	1	2	5	4	10	5	1	3	1	6	2	14	1	4	13	2	1	6
1	1	4	11	3	7	1	2	5	6	4	9	1	3	2	5	8	8	1	5	2	3	4	12
1	2	2	6	7	9	1	2	5	9	6	4	1	3	2	12	2	7	1	5	3	2	2	14
1	2	2	8	8	6	1	2	6	2	12	4	1	3	4	2	5	12	1	7	3	2	4	10
1	2	2	9	6	7	1	2	7	4	8	5	1	3	5	2	4	12	1	7	4	2	3	10
1	2	4	4	5	11	1	2	10	6	4	4	1	3	6	2	5	10						

127 — If the scale beams are of unequal length, the two weights are in inverse proportion to the length of the beams, a and b. Hence the ratio of the first weight to the real one equals $b \div a$, that of the second weight to the real one equals $a \div b$. Therefore, the real weight is the geometrical mean, or the square root of the product, of both wrong weights, in this case 143 ounces or 8 pounds 15 ounces.

128 – If the method previously explained is used, the number 421,052,631,578,947,368 will quickly be obtained.

129 – The general expression of the problem is 4 $(abcde)$ = $edcba$. It is evident that the looked-for number must begin with either 1 or 2, because otherwise the product would have six digits. The product, 4 $(abcde)$, must be an even number. Hence a can only be 2. $4e$, therefore, must end with a 2, so that e must be 3 or 8, and only 8 is eligible for the first digit of the product. Looking at the above expression of the problem, we may further conclude that b can be no more than 2 and must be an odd number. Hence $b = 1$. $4d$ must end on 8, so d must be either 2 or 7. It is easy to conclude that $d = 7$ and finally, $c = 9$. The looked-for figure is 21,978. This number is twice 10,989, the number which, when multiplied by 9, yields itself inverted. Amazingly, 10,989 has still other peculiarities. If we multiply it by 1, 2 . . . . 9, we get the following products:

10,989	×	1	=	10,989
10,989	×	2	=	21,978
10,989	×	3	=	32,967
10,989	×	4	=	43,956

10,989	×	5	=	54,945
10,989	×	6	=	65,934
10,989	×	7	=	76,923
10,989	×	8	=	87,912
10,989	×	9	=	98,901

Scanning these products we find that 10,989 × $(10 - a)$ is the inversion of 10,989 × a, a being 1, 2, 3 . . . . 9.

This peculiarity is in no way restricted to five-digit numbers. The four-digit number 1,089, the six-digit number 109,989, the seven-digit number 1,099,989, or quite generally, any number of the type 10,999 . . . . 99,989 has the same peculiarity; multiplied by $(10 - a)$ it yields the inversion of its multiplication by a.

Thus you can write down, for example, an 11-digit number— the only 11-digit number with which this is possible – the inversion of which equals itself multiplied by 9:

$$10,999,999,989 \times 9 = 98,999,999,901.$$

If you multiply this number by 2 you have 21,999,999,978, and if you multiply this number by 4 you get its own reversal, 87,999,999,912.

130 – If N is the number in question, e its unit and n the number of its digits, we can write the equation $e \times 10^{n-1} + \dfrac{N-e}{10} = \dfrac{N}{3}$ or $7N = 3e\,(10^n-1)$. It follows that either e equals 7 or (10^n-1) must be divisible by 7. e cannot be 7, because then N would be 3 (10^n-1), which implies that N has $(n + 1)$ digits instead of n, as presumed. Hence (10^n-1) must be divisible by 7. This can only be the case if n is a product of 6, that is, if $n = 6k$. If you try $k = 1$, you get as the smallest possible number $N = 3 \times e \times \dfrac{10^6-1}{7}$ $= 3 \times e \times 142{,}857$, in which e can be replaced only by 1 or 2. Thus you get $N = 428{,}571$ or $857{,}142$. By the way, we get the second solution for N by shifting the first two digits of the first solution to the end.

131 – The procedure is similar to the one shown in the previous problem. You have the equation $13N = 2 \times e \times (10^n-1)$, with $n = 6$. Hence $N = 2 \times e \times 76{,}923$. Since $\dfrac{N-e}{10} = \dfrac{153{,}845e}{10}$ is a five-digit integer, e can be only 2, 4 or 6. The three solutions are $N = 307{,}692$; $615{,}384$; $923{,}076$.

132 – You can write the equations $400 - x = 10y + 4$ and $400 + x = 400 + y$. Therefore, $x = y = 36$. The number is 364.

133 – The sum of all 12 possible combinations, $ab, ba, ac \ldots ,$ yields the equation $99a + 99b + 99c + 99d = 1000a + 100b + 10c + d$ or $901a = 89c - 98d - b$. Since neither c nor d can be bigger than 9, $901a$ cannot be bigger than 1,683, and thus a must be 1. From $901 = 89(c+d) + 9d - b$, it follows that $(c+d) = 10$ and $9d - b = 11$. Hence $d = 2$, $b = 7$ and the number in question is 1,782.

134 – If you call the number abc you can easily develop the equation $b = 8c - 10a$. c must be greater than 1 because otherwise $8c - 10a$ would yield a negative b. Now try 2, 3 .... 9 for c and by way of indefinite equations you will get the following results: 162, 243, 324, 405, 567, 648, 729.

135 – To solve this problem you have first to find out which primes divide into 70007. They are 7, 73 and 137. Three-digit products of these primes are 137, $7 \times 73 = 511$ and $7 \times 137 = 959$. You will get products of each of these numbers by adorning them with 7's at the beginning and the end.

136 – To obtain the first number you have the equation $27x = 100{,}000{,}001 + 10x$; $17x = 100{,}000{,}001$; $x = 5{,}882{,}353$. In the same way you will get the second number by dividing $1{,}000{,}000{,}001$ by 19; the result is $52{,}631{,}579$. The ten-digit number is obtained by dividing $100{,}000{,}000{,}001$ by 23; it is $4{,}347{,}826{,}087$.

137 – Numbers built like this are not primes, and neither are other numbers similarly built up. $1{,}001 = 11 \times 91$. $1{,}000{,}001 = 101 \times 9{,}901$. $1{,}000{,}000{,}001 = 1{,}001 \times 999{,}001$. The divisors of the number in the problem are 10,001 and 99,990,001. The general rule is: If the number of 0's framed by 1's equals a product of 3 increased by 2 – that is, $n = 3k + 2$ ($k = 0,1 \ldots$) – the two divisors may be written down as indicated.

138 – Any figure in which the sum of two neighboring digits does not exceed 9 will answer the requirement. For example, $263,542 \times 11 = 2,898,962$; $245,362 \times 11 = 2,698,982$. Work out the two multiplications the way you are used to doing and you will understand why.

139 – The only such pair is $18 \times 297 = 5,346$ and $27 \times 198 = 5,346$. In this instance, empirical search may be facilitated by the application of certain rules of divisibility.

140 – You have the following basic equations: $C + F - I = 0$ or 10; $B + E - H = 0$ or 9; $A + D - G = 1$ or 0; $A + B + C + D + E + F + G + H + I = 45$. Therefore, $G + H + I = 22$ or 13. Now it is easy to find a combination in accordance with the premises of the problem. For instance, $458 + 321 - 679 = 100$ or $257 + 189 - 346 = 100$.

141 – There is probably no theoretical equipment that can be used to solve this problem. You'll have to try all possible number combinations. The solution is $2^5 \times 9^2 = 2592$.

142 – The fraction has the general form $\dfrac{10x + y}{10y + z} = \dfrac{x}{z}$ or $9xz = y(10x - z)$. y must be 3 or a product of 3 because otherwise $(10x - z)$ must be divisible by 9, which is impossible except for the trivial case $x = z$. The fractions we are looking for are $\frac{26}{65}$, $\frac{16}{64}$, $\frac{19}{95}$, and $\frac{49}{98}$.

143 – There are three ways, namely 198, 199, 200, 201, 202; 28, 29 .... 51, 52; and 55, 56 .... 69, 70.

144 – It is best to start with the most difficult problem, the one with all nine numerals. If that one is fully understood, the solution of the five- and seven-digit number problems will be comparatively easy. Let us write the pattern of this problem this way:

$$a\ b\ c\ d\ e\ f\ g\ h\ i$$
$$-i\ h\ g\ f\ e\ d\ c\ b\ a$$

$\overline{A\ B\ C\ D\ E\ F\ G\ H\ I}$, with the capital letters of the difference representing the same nine different numerals as the small letters, though in a different order.

Simple reasoning indicates that a must be greater than 1 and E must be 9.

With regard to the sizes of corresponding numbers and number groups, four groups of solutions may be distinguished. The bold-faced digits or number combinations are bigger than the corresponding digits or combinations above or beneath. Group 1.

$$
\begin{array}{l}
\textbf{a}\,\textbf{b}\,\textbf{c}\,\textbf{d}\;e\;f\;g\;h\;i \\
-\,i\;h\;g\;f\;e\;\textbf{d}\,\textbf{c}\,\textbf{b}\,\textbf{a} \\
\hline
A\,B\,C\,D\,E\,F\,G\,H\,I
\end{array}
\quad \text{for instance:} \quad
\begin{array}{r}
9\,7\,8,4\,5\,\textbf{1},\textbf{3}\,\textbf{6}\,\textbf{2} \\
-\,2\,6\,3,1\,5\,4,8\,7\,9 \\
\hline
7\,1\,5,2\,9\,6,4\,8\,\textbf{3}
\end{array}
$$

We have the following relations:

$a - i = A$; $10 + i - a = I$; hence $A + I = 10$.
$b - h = B$; $10 + h - b - 1 = H$; hence $B + H = 9$.
$c - g = C$; $10 + g - c - 1 = G$; hence $C + G = 9$.
$d - f - 1 = D$; $10 + f - d - 1 = F$; hence $D + F = 8$.

Group II.

$$
\begin{array}{l}
\textbf{a}\,\textbf{b}\,c\,d\,e\,f\;g\;h\;i \\
-\,i\;h\;\textbf{g}\,f\,e\,\textbf{d}\,\textbf{c}\,b\,a \\
\hline
A\,B\,C\,D\,E\,F\,G\,H\,I
\end{array}
\quad \text{for instance:} \quad
\begin{array}{r}
6\,9\,3,8\,5\,2,\textbf{7}\,1\,4 \\
-\,4\,1\,7,2\,5\,8,\textbf{3}\,9\,6 \\
\hline
2\,7\,6,5\,9\,4,\textbf{3}\,1\,8
\end{array}
$$

$A + I = 10$; $B + H = 8$; $C + G = 9$; $D + F = 9$.

Group III.

$$
\begin{array}{l}
\textbf{a}\,b\,c\,\textbf{d}\;e\;f\;\textbf{g}\,\textbf{h}\;i \\
-\,i\;h\;\textbf{g}\,f\,e\,d\,c\,b\,a \\
\hline
A\,B\,C\,D\,E\,F\,G\,H\,I
\end{array}
\quad \text{for instance:} \quad
\begin{array}{r}
5\,1\,6,7\,4\,2,\textbf{9}\,\textbf{8}\,\textbf{3} \\
-\,3\,8\,9,2\,4\,7,6\,1\,5 \\
\hline
1\,2\,7,4\,9\,5,\textbf{3}\,\textbf{6}\,8
\end{array}
$$

$A + I = 9$; $B + H = 8$; $C + G = 10$; $D + F = 9$.

Group IV.

$$
\begin{array}{l}
\textbf{a}\;b\;c\,d\;e\;f\;g\,\textbf{h}\;i \\
-\,i\;\textbf{h}\;g\;f\,e\,\textbf{d}\,\textbf{c}\,b\,a \\
\hline
A\,B\,C\,D\,E\,F\,G\,H\,I
\end{array}
\quad \text{for instance:} \quad
\begin{array}{r}
7\,1\,8,9\,4\,6,\textbf{5}\,\textbf{3}\,2 \\
-\,2\,3\,5,6\,4\,9,8\,1\,7 \\
\hline
4\,8\,\textbf{3},2\,9\,6,7\,1\,\textbf{5}
\end{array}
$$

$A + I = 9$; $B + H = 9$; $C + G = 10$; $D + F = 8$.

To determine the difference, $AB \ldots I$: Let us take as an example group III, in which the sums of various pairs of numbers in the difference are 9, 8, 10 and 9. Among the numbers 1 to 9 we have to find four pairs yielding these sums. There are 7 different possibilities, namely, 1, 8; 2, 6; 3, 7; 4, 5; and (only giving the sequence $ABCD$) 1342; 2314; 2341; 3124; 4123; 4231. For each one of these possibilities 15 more permutations may be found by exchanging the number pairs, for instance, B and H. Thus for the first possibility, 1234, we find 1235, 8234, 8235, 1634, 1635, 8634, 8635, 1274, 1275, 8274, 8275, 1674, 1675, 8674, 8675.

To determine the number itself, $abcd \ldots$: This may be explained for the number combination $ABCD = 1274$. We may find $abcd$ in the following way: $a - i = A + 1$; $h - b = 9 - B$; $g - c = 10 - C$; $d - f = D + 1$. The four differences in these four equa-

144

tions equal 2, 7, 3 and 5. Thus we find the number pairs 5, 3; 1, 8; 6, 9; 7, 2.

In practically every case — with the exception of symmetrically arranged differences (see the previous paragraph!) such as 8, 6, 4, 2 — for every group of differences there is a complementary group. In our example it is 7, 5; 2, 9; 4, 1; 8, 3. If the differences are particularly favorably arranged, such as the group 1, 2, 4, 4, there may be 12 different number solutions, $a, b \ldots i$ of the problem for one single difference $A, B \ldots I$. On the other hand, often the numbers 1 to 9 are insufficient for the formation of the determined differences $a - i$ etc. and consequently, therefore, no number $a, b \ldots i$ can be found to fit the difference $A, B \ldots I$ in question. In any case, E is the remaining numeral not used for the formation of the differences, $a - i$, etc. Altogether, utilizing all possibilities referred to, there are 966 nine-digit numbers of the kind dealt with in this problem.

The seven-digit number problem is simpler than the nine-digit problem just treated because of its having two digits less, but on the other hand, it is complicated by the use of 0 and the lesser number of possible differences. Again we distinguish four distinct groups.

I.
$$\begin{array}{r} \mathbf{a}\,\mathbf{b}\,\mathbf{c}\,d\,e\,f\,g \\ -g\,f\,e\,d\,\mathbf{c}\,\mathbf{b}\,\mathbf{a} \\ \hline A\,B\,C\,9\,E\,F\,G \end{array}$$
for instance:
$$\begin{array}{r} 8,7\,9\,4,0\,3\,5 \\ -5,3\,0\,4,9\,7\,8 \\ \hline 3,4\,8\,9,0\,5\,7 \end{array}$$

$A + G = 10; B + F = 9; C + E = 8.$

II.
$$\begin{array}{r} \mathbf{a}\,b\,\mathbf{c}\,d\,e\,f\,g \\ -g\,\mathbf{f}\,e\,d\,\mathbf{c}\,\mathbf{b}\,a \\ \hline A\,B\,C\,9\,E\,F\,G \end{array}$$
for instance:
$$\begin{array}{r} 8,1\,7\,2,3\,9\,6 \\ -6,9\,3\,2,7\,1\,8 \\ \hline 1,2\,3\,9,6\,7\,8 \end{array}$$

$A + G = 9; B + F = 9; C + E = 9.$

III.
$$\begin{array}{r} \mathbf{a}\,\mathbf{b}\,c\,d\,e\,f\,g \\ -g\,f\,e\,d\,c\,\mathbf{b}\,\mathbf{a} \\ \hline A\,B\,C\,0\,E\,F\,G \end{array}$$
for instance:
$$\begin{array}{r} 2,9\,0\,4,5\,6\,1 \\ -1,6\,5\,4,0\,9\,2 \\ \hline 1,2\,5\,0,4\,6\,9 \end{array}$$

$A + G = 10; B + F = 8; C + E = 9.$

IV.
$$\begin{array}{r} \mathbf{a}\,b\,c\,d\,e\,f\,g \\ -g\,\mathbf{f}\,e\,d\,c\,b\,\mathbf{a} \\ \hline A\,B\,C\,0\,E\,F\,G \end{array}$$
for instance:
$$\begin{array}{r} 6,1\,0\,7,8\,3\,2 \\ -2,3\,8\,7,0\,1\,6 \\ \hline 3,7\,2\,0,8\,1\,6 \end{array}$$

$A + G = 9; B + F = 8; C + E = 10.$

Altogether, there are 157 solutions to this problem.

As to the problem of the five-digit number, there are only 8 solutions, 58,923, 60,273, 60,732, 69,723, 70,254, 76,941, 89,604 and 96,732. Of course, 0 at the beginning or end of the number is out of the question.

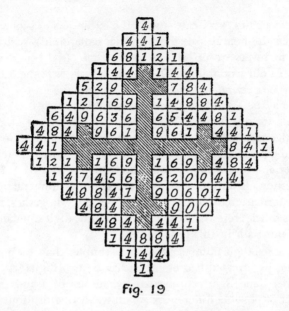

Fig. 19

Other solutions with slightly different numbers are possible.

146 – The sum of a four-digit number and its inversion is always divisible by 11, because $1000a + 100b + 10c + d + 1000d + 100c + 10b + a = 1001 (a + d) + 110 (b + c)$. Now the question arises whether the divisibility by 11 of the sum of two squares doesn't presuppose the divisibility by 11 of each single square. The answer is yes, for the following reasons: Every number divided by 11 has the remainder 0 or ± 1 or ± 2 or ± 3 or ± 4 or ± 5. Consequently, its square, divided by 11, must have one of the remainders, 0, 1, 4, 9, 5 or 3. If the sum of two squares is divisible by 11, there must be among these remainders one which, if added to itself or one of the others, yields 11 or a product of 11. Only the remainder 0 satisfies this requirement, and that implies that each of the two squares (and in this particular case their roots also) are divisible by 11. The square in question can end only on 1, 4, 5, 6 or 9, because only these numbers occur as units of the squares of 1, 2 . . . . 9. The unit of the square cannot be 5, because then the two last numerals would have to be 25, which implies that the inversion would have to begin with

52. But there is no square of an integer between 5200 and 5299. Consequently, the square you are looking for is somewhere in the intervals 1000 to 2000, 4000 to 5000, 6000 to 7000, 9000 to 10,000 and hence its root lies between 31 and 45, 63 and 71, 77 and 84, and 94 and 100. Since the root must be divisible by 11, only 33, 44, 66 and 99 would be eligible. Now it is easy to ascertain the right square: $1089 = 33^2$, $9801 = 99^2$.

147 — The squares in question can only end on 00 or 4 (see solution of the previous problem), because a square ending on 6 would have an odd digit next to the last. The only available square ending on 00 is 6400. The root of a square ending on 4 has the form $10\,a \pm 2$ so that only the figures 32, 38, 42, 48 . . . . 92, 98 are available. However, only $68^2 = 4624$, $78^2 = 6084$ and $92^2 = 8464$ satisfy the premise of the problem.

148 — You can write the equation $100x - y = (x - y)^2$. You can conclude that x is an integer if $(2500 - 99y)$ is a square, S^2. From this you may develop the equation $99y = (50 - Q) \times (50 - Q)$. Now you replace y by uv. 99 being either 1×99 or 3×33 or 9×11, you have the following possibilities: $u = 50 \pm Q$, $(99v = 50 \pm Q)$; $3u = 50 \pm Q$, $(33v = 50 \pm Q)$ and $9u = 50 \pm Q$, $(11v = 50 \pm Q)$. The rest is easy. From the three pairs of equations you will finally get the following three numbers which satisfy the conditions of the problem: $9801 = (98 + 01)^2$; $3025 = (30 + 25)^2$; and $2025 = (20 + 25)^2$.

149 — From $(100x - 10y - z)^2 = 10,000a - 1,000b - 100c - 10d - e$ and $(100z - 10y - x)^2 = 10,000e - 1,000d - 100c - 10b - a$, it follows that $x^2 = a$, $2xy = b$, $xz = c$, $2yz = d$ and $z^2 = e$. Since the eight letters represent one-digit numerals, we can empirically find the following solutions and their inversions:

x	y	z	a	b	c	d	e
1	0	2	1	0	4	0	4
1	0	3	1	0	6	0	9
1	1	2	1	2	5	4	4
1	1	3	1	2	7	6	9
1	2	2	1	4	8	8	4

150 — Division of any number by 9, leaves a remainder of 0, ± 1, ± 2, ± 3 or ± 4. Consequently, division of any square number by 9 can leave only the remainder 0, 1, 4 or 7, because the squares of the above numbers are 0, 1, 4, 9 and 16, of which 9 is divisible into 9 and 16 leaves the remainder 7. Perhaps you remember how to find out the remainder of a division by 9. You add up all the digits, leaving the 9's out, and divide by 9, do the same thing with the result, and so on, until a remainder smaller than 9 is left. If we proceed in this way with the number given in our problem we end up with the numeral 5. This is not one of the four numbers, 0, 1, 4 and 7, which single out any square number. Consequently, the number is not a square.

151 – If we denote the second and fourth from the end digits by a, the last digit by b and the digit third from the end by $(b + 1)$, we have the equation $1010a + 101b = x^2 - 100$, or $101(10a + b) = (x + 10)(x - 10)$. 101 must equal $(x + 10)$ and $x = 91$, thus $x^2 = 8281$.

152 – This problem requires the proof that any number $N = 4444\ldots.44448888\ldots..8889$, that is, a number with n 4's and $(n - 1)$ 8's and a 9 as unit is always a square: $9N = 4444\ldots.. 44448888\ldots..8889 \times 9 = 40000\ldots..00040000\ldots..0001 = 4 \times 10^{2n} + 4 \times 10^n + 1 = (2 \times 10^n + 1)^2$. $N = \left(\dfrac{2 \times 10^n + 1}{3}\right)^2$, which implies that N is always a square.

16 has the same peculiarity as 49, the numbers 1156, 111556, 11115556, and so on being squares.

153 – The next to the last digit of the root can only be 4, because with 3, say, $39^2 = 1521$, the first two digits are less than the smallest two-digit square, 16; and on the other hand, with 5, $(50 + x)^2 = 2500 + 100x + x^2$, which would in any case raise the square 25, but never enough to reach 36. The unit of the root can only be 1, because otherwise the section of the square $(2 \times 40y + y^2)$ would exceed 100 and thus add to the square 16. Therefore, the square number you are looking for is $1681 = 41^2$.

154 – If you denote the sum of the squares of the numbers along any side of the triangle by S and keep in mind that the squares of the numbers at the corners occur in both adjacent sides, and that the sum of all the squares from 1 to 9 is 285, you will find that the squares of the numbers at the corners add up to $3S - 285$ or $3(S - 95)$, that is, to a multiple of 3. Consequently, for the three corners the only numbers eligible are those with squares which add up to a multiple of 3. The remaining numbers have to be inserted along the sides of the triangle in such a way that the sums of the squares along each side amount to 95 – one-third of the sum of the corner squares. You will soon find that only 2, 5 and 8 are eligible for the corners and that the numbers have to be arranged around the triangle like this (Fig. 20):

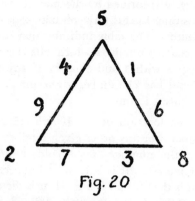

Fig. 20

Amazingly, the sums of the numbers along each side of the triangle are also equal, namely 20.

155 — If x is the first number of a series of consecutive integers and y is the number of the integers in the series, then the last number of this particular arithmetical progression is $x + y - 1$. Using the formula for the sum of an arithmetical progression you will find the algebraic formulation of the problem to be $\frac{1}{2}y(2x + y - 1) = x(x + y - 1)$, from which you can derive the equation $2x = 1 \pm \sqrt{(y - 1)^2 - y^2}$. Using the formula for the sum of an arithmetical progression you will find the algebraic formulation of the problem to be $\frac{1}{2}y(2x + y - 1) = x(x + y - 1)$, from which you can derive the equation $2x = 1 \pm \sqrt{(y - 1)^2 - y^2}$. Using the laws governing Pythagorean number triples, $y = a^2 - b^2$, $y \pm 1 = 2ab$ and $z = a^2 + b^2$, you will get $y \pm 1 = a^2 - b^2 \pm 1 = 2ab$, or $a^2 - 2ab - b^2 = \pm 1$. If $2b^2$ is added to both sides of the equation, $(a - b)^2 = 2b^2 \pm 1 = u^2$ will result, where u is an auxiliary. This yields $b = \frac{1}{2}(u^2 \pm 1)$. From $(a - b)^2 = u^2$ follows $a = u + b$. Thus, b is an integer if $\frac{1}{2}(u^2 \pm 1)$ is a square. Now it is easy empirically to find values for b. The table shows a few of the possible values for u, b, a, y and x.

u	b	a	y	x
1	1	2	4	3
3	2	5	21	15
7	5	12	120	85
17	12	29	697	493
41	29	70	4060	2871
99	70	169	etc.	etc.

There is a certain regularity governing the numbers of the table, which enables you to fill in higher values with the aid of the first lines and without further figuring. a in one line always equals b

149

in the line below. *u* is formed by the sum of the *b*'s in that line and the previous one. In this way you may continue building up the table indefinitely. The table indicates that the first section of a series of consecutive numbers which fills the requirements of the problem begins with 3 and consists of 4 numbers, the next begins with 15 and has 21 numbers. The number of such arithmetical progressions is infinite.

156 — To find the location of the 1000th column you have to divide this number — or any other given number, for that matter — by 12 (twice the number of the columns minus two) and determine the remainder. If it is 1, the column in question is the first; if 0 or 2, it is the second; if 3 or 11, it is the third; if 4 or 10, it is the fourth; if 5 or 9, it is the fifth; if 6 or 8, it is the sixth; if 7, it is the seventh column. The slave finished his march — at least, that's what we hope — at the fourth column.

157 — Under this system, the number of a male ancestor equals twice the number of his son or daughter; the number of a female ancestor equals twice the number plus 1. Therefore, we can write down 1 for the contemporary in New Orleans, 2 for his father, 5 for that man's mother, 11 for that woman's mother, 22 for that woman's father, 44 for that man's father, 89 for that man's mother, 179 for that woman's mother, and so on.

If you run the number scale backward, starting at 1000, continually dividing by 2 and watching the remainders, you will find the following relationship for the obscure ancestor number 1000: he is the father of the father of the father of the mother of the father of the mother of the mother of the mother of the mother of our contemporary. (These are the figures you will get by division, *r* meaning remainder: 1000, 500, 250, 125, *r*, 62, 31, *r*, 15, *r*, 7, *r*, 3, *r*, 1.)

158 — One third of those who flunked did not fail in English, one fourth did not fail in French and one fifth did not fail in Spanish. Therefore, it is possible that $\frac{47}{60}$ of all who flunked were not among those who failed in all three subjects. This means that at least $\frac{13}{60}$ were. Since this fraction of all those who flunked was stated to be 26, altogether 120 must have failed to pass the exam.

159 — We have four equations for four unknowns, namely $4x = 5y = 6z = 7u = x + y + z + u + 101$. From these equations we find *x*, *y*, *z* and *u* to be 105, 84, 70 and 60, respectively, which are the amounts in dollars Fred has to wager on each of the horses, in the order of the odds given in the problem.

160 — If we denote the number of the members of the family without the twins by x and the number of days the crate of oranges lasted before the arrival of the twins, by y, we have the equations $xy = (x + 2)\ (y - 3) = (x + 6)\ (y - 7)$, which divulge that the family consisted of 12 mouths (with the parents but without the twins, a number which, by the way, equals the number of children in that family) and that the crate lasted 21 days before the twins joined their brothers and sisters.

161 — After the distribution of the bread each had $\frac{5}{3}$ loaves. One therefore, had contributed $\frac{4}{3}$ loaves to hobo number three's share, the other, $\frac{1}{3}$. Hence the third hobo had to divide his money in the ratio 4 : 1, or 80 cents and 20 cents.

162 — We can write down the following equations:

$$\frac{x}{2} + \frac{n}{3} = \frac{x + y + z}{2}$$

$$\frac{2y}{3} + \frac{n}{3} = \frac{x + y + z}{3}$$

$$\frac{5z}{6} + \frac{n}{3} = \frac{x + y + z}{6}$$

where x, y and z are the shares the three men took from the pile and n is what all three of them together returned, namely, $n = \frac{x}{2} - \frac{y}{3} - \frac{z}{6}$. These four basic equations with four unknowns yield only two equations for x, y and z; $y = 13z$ and $x = 33z$. These two indefinite equations, however, allow us to determine the smallest number of coins, $x + y + z$, with which the described transaction can work, 282 ducats. From the pile formed by that sum the three men took $x = 198$, $y = 78$ and $z = 6$ ducats. Their legitimate shares were 141, 94 and 47 ducats.

163 — Since both birds reach the milestone simultaneously it must have the same distance, a, from both spires. With the aid of Pythagoras' theorem you get the equations $a = 30^2 + x^2 = 40^2 + (50 - x)^2$, x representing the distance of the milestone from one of the towers. You will find the distance from the towers to be 32 and 18 rods, respectively.

164 — The most unfavorable case as to 0's would be four numbers with two 0's each. Hence eight 0's are required. The most unfavorable case as to 1's would be hymn 111 and three more ending on 11,

and so on. Therefore, the following numbers of plates are necessary: eight 0's, nine 1's, nine 2's, nine 3's, nine 4's, nine 5's, twelve 6's, eight 7's and eight 8's; altogether 81 plates.

165 – We hope you have remembered that the ratio of weekdays and Sundays is not the same every year, a fact which will enable you to solve the problem. There are four different combinations of weekdays and Sundays in a year. Any ordinary year has either 313 weekdays and 52 Sundays or 312 weekdays and 53 Sundays, while a leap year has either 314 weekdays and 52 Sundays or 313 weekdays and 53 Sundays.

If we denote the two rates of distribution by x and y, we have for the first of the above four cases the Diophantine equation $365x = 313y + 1$, which yields as the smallest possible rates $307 and $358 and a total fortune of $112,055. Investigation of the second case yields the rates of $62 and $53 and a sum total of $19,-345. The third case has no solution. The fourth case yields $189 and $221 for the distribution rates and a total fortune of $69,174. Therefore, the year we endeavor to find (case 2 because of the minimum-fortune requirement) must be an ordinary year with 53 Sundays. 1933 was such a year, and therefore, the philanthropist took his resolution in 1932.

166 – If we designate McMillan's salary with x and his disbursements with y, he saves $x - y$, while Barber saves $\frac{2x}{3} - \frac{y}{2}$. Now McMillan saves only half as much as Barber; therefore, this expression equals $2(x - y)$. It is now easy to find that the ratio of McMillan's disbursements to his wages is 8 : 9.

167 – Let's call the amounts to be paid on the first and second days x and y. This is what the pirate chief had to deliver on the ten days: x, y, $x + 2y$, $3x + 5y$, $8x + 13y$, $21x + 34y$, $55x + 89y$, $144x + 233y$, $377x + 610y$, $987x + 1597y$, the sum of which ought to be 1,000,000. By way of a Diophantine equation we get as the first rate 144 pieces of silver and as the second rate 298 pieces.

168 – As with all puzzles of this kind you have to figure backward from the end. You will soon find out that 160 passengers left New York.

169 – Altogether 207 votes were cast, first apparently 115 in favor and 92 against, then 103 in favor and 104 against.

170 – The ratio of the male to the female inhabitants of the little republic is 2 : 3.

152

171 – If you subtract 52 from 17,380 and divide by 57 you will get 304. The correct divisor is 364 and hence the original dividend was 20,800.

172 – The probability that the father did not receive the letter is 1 in 6. The ratio of the answers received to the letters sent is 25 : 36.

173 – You do not need algebraic formulas for this one. Just get a piece of paper and a pencil and start from the end, that is, from the 128 marbles in each basket. That way you will quickly find that the baskets contained 449, 225, 113, 57, 29, 15 and 8 marbles. Solution of problem No. 168 was found by a similar procedure.

174 – If you make a cut all through the paper wound on the spool, parallel to the spool, and unfold the cylindrical paper mass, the resulting figure will be a trapezium, the two parallel sides of which measure 16π and 4π inches and the altitude of which is $\dfrac{16-4}{2} = 6$ inches. If you divide the area of the trapezium $\dfrac{6 \times 16 + 4}{2\pi}$ square inches, by the thickness of the paper, $\frac{1}{250}$ inch, you have the length of the paper strip. It is $15,000\pi$ inches or $416\frac{2}{3}\pi$ yards, that is approximately 1,309 yards.

175 – Let a and b be the two women, and I the little son of a and A, II the son of b and B. Since both sons are sons of the sons of the women, a and b, and since A cannot be the son of his wife,

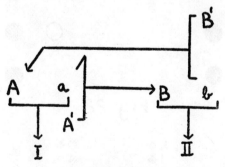

Fig. 21

a, A must be the son of b, B the son of a, both offspring of previous marriages of a and b with A' and B'. Hence I is the son of a and the grandson of b, II is the son of b and the grandson of a. II is also the half-brother of A (husband of the woman a) because both II and *A* have the same mother. And thus the statement of the two XIIIth-century women was correct. (See Fig. 21.)

176 —

```
*  *  W  B  W  B  W  B  W  B  W  B
B  W  W  B  W  B  W  B  W  *  *  B
B  W  W  B  *  *  W  B  W  W  B  B
B  W  W  B  B  W  W  *  *  W  B  B
B  *  *  B  B  W  W  W  W  W  B  B
B  B  B  B  B  W  W  W  W  W  *  *
```

By the way, it is to be noted that the second and fifth pieces are not moved at all.

177 —

If you have a greater number of coins you place the fourth on the first, then the sixth on the second, and so forth, until only the eight inner coins remain. Then you continue as shown below.

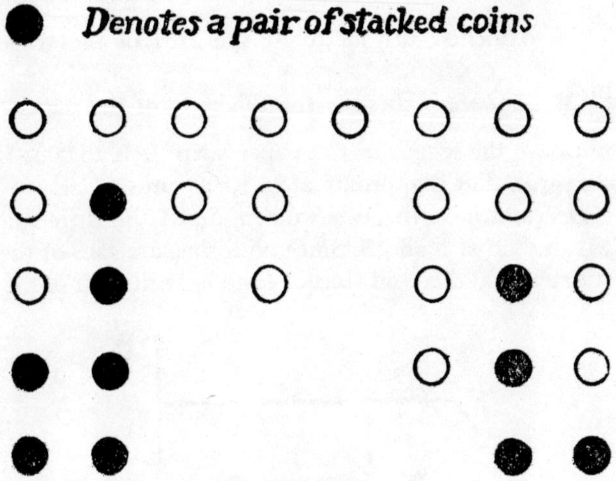

Denotes a pair of stacked coins

Fig. 22

178 —

A♡	K♢	Q♣	J♠
Q♠	J♣	A♢	K♡
K♣	A♠	J♡	Q♢
J♢	Q♡	K♠	A♣

179 — If you denote the sum of all ten numbers by S and the sum of the four numbers along any of the five lines of the pentagram by s, you get the formula $5s = 2S$, because if you add the five s's each number occurs twice. In the problem, 24 is given as the sum,

s, and indeed, 24 is the smallest number for this problem.

You begin by writing the five smallest numbers into the circles of the inner pentagon of the figure and then try to discover a method for finding out the numbers pertaining to the five points of the pentagram. Let us start by calling x the unknown number at the point G. Then the sum of the numbers along the two lines intersecting at G is $2s - x$. Now all numbers with the exception of F, H and D are accounted for. Moreover, we can state that $F + H = s - (A + B)$ and, consequently, $2s - x + s - (A + B) + D = S = \dfrac{5}{2}$. Hence $x = \frac{1}{2}s + D - (A + B)$.

This implies that the number at any of the five points equals one half the sum of the numbers along any line (s or, in this case, 24), plus the number of the pentagon which is diametrically opposite, minus the two numbers of the pentagon closest to the unknown number. The numbers for the intersections of the pentagram ascertained with the aid of this procedure are shown in Fig. 23.

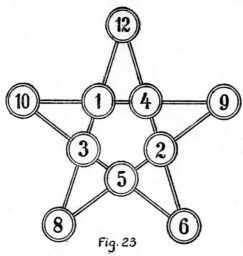

Fig. 23

As mentioned in the problem, you are free to choose any five numbers for the corners of the inner pentagon, as well as the sum of the four numbers along the lines. However, as also mentioned, it might happen very often that numbers occur more than once and that you cannot do without negative numbers.

180 — There are 30 different ways of distributing the six colors. You may reason this way: Any of the six faces of a die may be painted with the color A. For the opposite face five possibilities remain, B, C, D, E and F. When you have chosen one of these colors, four colors remain, C, D, E and F. One of the four remaining "vertical" faces must be painted with the color C. Then three possibilities remain for coloring the opposite face, D, E and F. When you have decided which to choose, there still remain two ways of distributing the two remaining colors. Thus, you have $5 \times 3 \times 2 = 30$ variants of distribution of the six colors over the six faces of a die.

181 — You may use either 27 or 29.

182 — The equation $22^2 + 19^2 = 13^2 + 26^2$ can be written $19^2 - 13^2 = 26^2 - 22^2 = 192$. Now, the difference of two squares equals the product of the sum and the difference of their roots $(a^2 - b^2 = (a + b)\ (a - b))$. We now express 192 in five different pairs of even factors: 2×96, 4×48, 6×32, 8×24 and 12×16. Each factor divided by two yields the products 1×48, 2×24, 3×16, 4×12 and 6×8. The sums and differences of these numbers are 47 and 49, 22 and 26, 13 and 19, 6 and 16, and 2 and 14. These are the figures for which you were looking and you will see that four of them are already inserted in the tablets. The correct clockwise arrangement of the ten numbers is 22, 19, 6, 14, 47, 26, 13, 16, 2 and 49.

183 — If you consider only the sums along the sides of the triangle AHK ($= 3 \times 26$), you will realize that the numbers at A, H and K are counted double. Since the sum of the corners of the inner hexagon is also 26, you will find that $A + H + K = B + E + L = 26$. From $A + H + K = H + I + J + K$, it follows that $A = I + J$, which means that the sum of two neighboring corners of the hexagon equals the number at the opposite triangle corner. This indicates that the numbers 11 and 12 cannot occur in the hexagon because otherwise numbers greater than 12 would turn up at the triangle corners, and this possibility is excluded by the premises of the problem.

If you investigate which number triples would yield a sum of 26 you will find the following:

(1)	12,	11,	3	(5)	11,	10,	5
(2)	12,	10,	4	(6)	11,	9,	6
(3)	12,	9,	5	(7)	11,	8,	7
(4)	12,	8,	6	(8)	10,	9,	7

Among these number triples you will have to pick two for the two triangles, and the numbers 12 and 11 will have to occur in

156

these two number triples. This means that if duplications of numbers are to be avoided, only the combinations (1, 8), (2, 6), (2, 7), (3, 7) and (4, 5) can be considered.

You may begin with the first of these possible combinations, (1, 8), and arrange the numbers 12, 3 and 11 clockwise at the corners of one of the triangles. A rearrangement of these three numbers cannot yield anything new since the first arrangement can always be obtained by a rotation or a reflection. For F and C only the number pair 1, 2 or 2, 1 can be used. It follows from the equations $A = I + J, L = C + D$ and $C + D + G + J + I + F = 26$ that $A + F + L + G = 26$. This indicates that every four points forming a rhombus must also add up to the sum of 26. For the point L, 10, 9 or 7 has to be considered. If M were 10 and F were 1, G would be 3, a number which is no longer available. The other case, $F = 2$, is also out of the question. A similar line of reasoning excludes $M = 9$. Hence only 7 remains for L. Counting clockwise from F, we get for the four upper numbers of the hexagon, 1, 2, 5 and 6 or 2, 1, 6 and 5. Now only the figures 4 and 8, or 8 and 4, are to be placed at I and J. Neither with $F = 1$ nor $F = 2$, can J be 8, because in that case you would have two adjacent numbers with a sum greater than 12. Thus, the two solutions based on the number triples (1, 8) are:

A	B	C	D	E	F	G	H	I	J	K	L
12	10	2	5	9	1	6	11	8	4	3	7
12	9	1	6	10	2	5	11	8	4	3	7

For the combinations (2, 6), (3, 7) and (4, 5) you can easily find two primary arrangements which, by rotation and reflection, may be transformed into further variations. It is not difficult to prove that the combination (2, 7) doesn't work.

184 — First, place 5 at the uppermost point of the star. Then arrange four numbers on the horizontal line below, in such a way that the sum of the two outside numbers is 10, and of the two inner ones, 20, and that the difference between the two outside numbers is twice the difference between the two inner numbers (7, 11, 9 and 3). Starting from these basic numbers, insert numbers which yield the sum 15 in the relative positions indicated by the dotted lines. It is not difficult finally to find the right spots for the remaining numbers, 13, 2, 14 and 1. In this basic solution, 13 may be exchanged with 1, and 14 with 2. In both these arrangements, moreover, for every number its difference from

15 may be substituted. Altogether, 56 solutions to the problem have been found through such exchanges so far.

185 — The distribution of the pieces has to take place in this order:

2	2	2	2	0		0	1	1	1	5
2	2	0	3	1		0	1	1	0	6
3	3	0	0	2		0	0	2	0	6
3	0	1	1	3		0	0	0	1	7
3	0	0	2	3		0	0	0	0	8
4	0	0	0	4						

This sequence of moves represents the only possible solution of the four-hole Tchuka Ruma. There is no possible solution with 6 holes and a Ruma hole, and with 3 pieces in each hole. But the more complicated variant of 8 holes and a Ruma hole, and 4 pieces in each hole, has a solution.

186 — For the game played by three people the score for player A can be figured with the help of the formula, score $= 2sA - (sB + sC)$, and accordingly, the formulas for B and C are $2sB - (sA + sC)$ and $2sC - (sA + sB)$. There is an even simpler way of solving the problem. If $S = sA + sB + sC$, the three scores are $3sA - S$, $3sB - S$ and $3sC - S$.

For four players you proceed this way: $S = sA + sB + sC + sD$, which in our example equals 1,110 cents. This sum was won as well as lost (by other players). A's winnings were sA cents, his losses were half of the winnings of the other players with the exception of those won by any of his partners jointly with himself. Thus, his losses were $\dfrac{S - 2sA}{2}$ and his score is $sA - \dfrac{S - 2sA}{2}$ or $2sA - \dfrac{S}{2}$. Corresponding formulas may serve for quickly computing the scores of the other players. This system of accounting can be used in bridge.

187 — If you denote all the cards lying on the table by $2n$, you have the equation $ax(a + 2) = 10n + 4$, where a is the length of the table and $10n$ is the product of the number of cards divided by the area of one card. If you add 1 to both sides of the equation you get $a^2 + 2a + 1 = 10n + 5$, that is, a square number ending in 5. Only 225 need be considered. Then $2n = 44$, which means that 8 cards are left over, 4 hearts and 4 clubs. Consequently, 9 clubs are lying on the table.

188 — If you have figured back from the six equal fortunes of the gamblers at the end of the game you will have found that they started the night with 193, 97, 49, 25, 13 and 7 dollars. You may even have found empirically that these figures are the result of the formula $6 \times 2^n + 1$, n being the numbers 0 to 5.

189 – 13 games. One of the men had won 3, the other 10.

190 – You can figure the stakes with the aid of the three equations, $36x - x - y - 2z = 18y - y - x - 2z = 6z - x - y - 2z = 2,100$. $x = 100$, $y = 200$, $z = 600$.

191 – Mary had 4 and 7 = 11; Betty, 1 and 6 = 7; Henry, 2 and 3 = 5; Charlie, 5 and 9 = 14; and Mr. Smith, 8 and 10 = 18.

192 – At the beginning the pool was $765. The first winner took $191; the second, $143; the third, $107; and the fourth, $80. $240 remained. Thus, the four gamblers went home with $251, $203, $167 and $140.

193 – You can win – or rather must win unless you make a mistake – if you can reach 7, 12 or 17. You will be sure to reach one of these marks only if you start with 1, though there remains a slight chance if you begin with 2. In all other cases, your friend can easily reach 7, 12 or 17, and he may win.

194 – Figuring backward from the last dollar, it is easy to find out that Earl once owned $61.

195 – If the husbands are A, B, C and D and their respective wives, a, b, c and d, you have this distribution:

	First Table	Second Table
First evening:	Ad against Bc	Da against Cb
Second evening:	Ac against Db	Ca against Bd
Third evening:	Ab against Cd	Ba against Dc

If you like this problem, you may try your ingenuity on a variant with 8 couples, 4 tables and 7 days.

196 – Any of the wives has won in x games $\$x^2$, her husband has lost in y games $\$y^2$, x and y being integers. As a couple they won $\$200 = x^2 - y^2 = (x - y)(x - y)$. Thus, the gist of the problem is to split 200 into a product of two even numbers. This can be done in three different ways, 2×100, 4×50, and 10×20. Hence you get three solutions, one for each of the couples: $x = 51$, $y = 49$; $x = 27$, $y = 23$; and $x = 15$, $y = 5$. Therefore, the three wives won $2,601, $729 and $225, respectively, their husbands lost $2,401, $529 and $25, in the same order. Considering the statements at the end of the problem, these figures indicate that Otto is married to Beatrice, Eric to Ann, and John to Margie.